# A PHOTOGRAPHIC ATLAS

# OF THE HUMAN BODY

## With Selected Cat, Sheep, and Cow Dissections

GERARD J. TORTORA

*Bergen Community College*

JOHN WILEY & SONS, INC.

New York • Chichester • Weinheim • Brisbane • Singapore • Toronto

Printer and Binder: Von Hoffmann Press

This book was printed and bound by Von Hoffmann Press, Inc. The cover was printed by Von Hoffmann Press, Inc.

The paper in this book was manufactured by a mill whose forest management programs include sustained yield harvesting of its timberlands. Sustained yield harvesting principles ensure that the number of trees cut each year does not exceed the amount of new growth.

This book is printed on acid-free paper ∞

*Library of Congress Cataloging in Publication Data:*
Tortora, Gerard J.
    A photographic atlas of the human body : with selected cat, sheep, and cow dissections /
Gerard J. Tortora.
        p.  cm.
    ISBN 0-471-37487-3
    1. Human anatomy Atlases.   2. Cats—Dissection.   3. Sheep—Dissection.   4. Cows—Dissection.   I. Title.
    [DNLM: 1. Anatomy Atlases. 2. Dissection Atlases. 3. Physiology Atlases. / QS 17 T712p 2000]
    QM25.T67   2000
    611'.022'2—dc21
    DNLM/DLC
    for Library of Congress                                                                99-32135
                                                                                                CIP

ISBN: 0-471-37487-3

Printed in the United States of America.

10  9  8  7  6  5  4  3  2

# TABLE OF CONTENTS

*A Photographic Atlas of the Human Body with Selected Cat, Sheep, and Cow Dissections* is designed to accompany any textbook of anatomy or anatomy and physiology and may be used in conjunction with or in lieu of a laboratory manual.

The study of the gross anatomical features of the human body is enhanced by the use of a photographic atlas to supplement your experience in the dissecting room or in those courses that do not include actual dissection. The clearly labeled cadaver photographs in this atlas were provided mainly by Mark Nielsen, of the University of Utah. They are organized by body system, and provide you with a stunning, visual reference to gross anatomy. The cadaver photos are supplemented by histological aspects of various organ and surface anatomy photos. You will benefit further from the helpful orientation diagrams, which accompany many of the photographs.

As you will see from the table of contents, this atlas covers all of the topics discussed in a typical anatomy or a combined anatomy and physiology course. The atlas begins with anatomical orientation and tissues, then progresses through each organ system and ends with surface anatomy. Where comparisons are helpful sheep and cow dissection photos appear with their human counterpart. A complete set of cat dissection photographs appears in an appendix. In addition, a useful correlation guide between the atlas

and A.D.A.M.® Interactive Anatomy points students who have access to that software to corresponding anatomical structures in the software.

Each photomicrograph in this atlas is accompanied by a line diagram and an inset that indicates a primary location of the tissue in the body. An icon with each photograph indicates whether it is a light or an electron micrograph and gives its magnification.

## Acknowledgments

There are many people whom I wish to thank for their efforts in helping me prepare this atlas. Claire Brassert, Developmental Editor, guided the atlas to completion. Her professionalism and editorial judgments have earned my respect. Sharon Montooth, Production Supervisor, expertly and efficiently moved the atlas through the production process. Each time I work with Sharon, the experience becomes more rewarding. Bradley Burch, Photo Coordinator, supervised the photography of the animal photos in the atlas. Andrew Ogus, Text and Cover Designer, designed the striking cover which encapsulates the contents of the atlas. I would also like to thank Dell Redding, of Evergreen Valley State College, Dennis Strete, of McLennan Community College, and Patricia Munn, of Longview Community College, for their contributions to the development of the atlas.

I would like to thank the following people for their opinions and advice. These instructors answered surveys, attended focus groups, or reviewed manuscript in the preparation of the atlas:


William Cliff, Niagra University
John Conroy, University of Winnipeg
James Crowder, Brookdale Community College
Brian Curry, Grand Valley State College
Pat Dementi, Randolph-Macon College
Deanna Ferguson, Jackson State Community College
Greg Garman, Centralia College
Keith Graham, Lutheran College of Health Professions
Charles Grossman, Xavier University
Ann Harmer, Orange Coast College
Sue Higgins, Medicine Hat College
Jane Horlings, Saddleback Community College
Bill Horrine, Alvin Community College
Earl Lindbergh, Davidson College
Patricia Mansfield, Santa Ana College
John Martin, Clark College
Javanika Mody, Anne Arundel Community College
Bill Montgomery, Charles County Community College
J. Stephen Noe, Ivy Tech State College
Nathan Norris, West Valley Community College
Michael Palladino, Brookdale Community College
Michael Patrick, Pennsylvania State University

Dan Porter, Amarillo College
Rebecca A. Pyles, East Tennessee State University
Penny Revelle, Essex Community College
Jackie Reynolds, Richland College
Wayne Seifert, Brookhaven College
Rhonda Shepperd, College of West Virginia
Marvin Sigal, Gaston College
David Skyrja, UWC-Waukesha
David Smith, San Antonio Community College
Sandra Stewart, Vincennes University
Sarah Strong, Austin Community College
Kristin Stuempfle, Gettysburg College
Claudia Williams, Campbell University
Ned Williams, Mankato State University
E.J. Zalisko, Blackburn College


In preparing this photographic atlas, I have strived to keep uppermost in my mind your needs, as a student. As always, I could benefit from your comments and suggestions, which I hope you will send to me at the address below.


Gerard J. Tortora
Science and Health, S229
Bergen Community College
400 Paramus Road
Paramus, NJ 07652

T A B L E  1 . 1 | **Directional Terms**

| Name | Definition | Example |
|---|---|---|
| **Superior** (soo′-PEER-ē-or) (**cephalic** or **cranial**) | Toward the head or the upper part of a structure. | The heart is superior to the liver. |
| **Inferior** (in′-FEER-ē-or) (**caudal**) | Away from the head or toward the lower part of a structure. | The stomach is inferior to the lungs. |
| **Anterior** (an-TEER-ē-or) (**ventral**)[*] | Nearer to or at the front of the body. | The sternum (breastbone) is anterior to the heart. |
| **Posterior** (pos-TEER-ē-or) (**dorsal**)[*] | Nearer to or at the back of the body. | The esophagus is posterior to the trachea. |
| **Medial** (MĒ-dē-al) | Nearer to the midline[†] or midsagittal plane. | The ulna is on the medial side of the forearm. |
| **Lateral** (LAT-er-al) | Farther from the midline or midsagittal plane. | The lungs are lateral to the heart. |
| **Intermediate** (in′-ter-MĒ-dē-at) | Between two structures. | The transverse colon is intermediate to the ascending and descending colons. |
| **Ipsilateral** (ip-si-LAT-er-al) | On the same side of the midline or midsagittal plane. | The gallbladder and ascending colon are ipsilateral. |
| **Contralateral** (CON-tra-lat-er-al) | On the opposite side of the midline or midsagittal plane. | The ascending and descending colons are contralateral. |
| **Proximal** (PROK-si-mal) | Nearer to the attachment of a limb to the trunk; nearer to the point of origin. | The humerus is proximal to the radius. |
| **Distal** (DIS-tai) | Farther from the attachment of a limb to the trunk; farther from the point of origin. | The phalanges are distal to the carpals. |
| **Superficial** (soo′-per-FISH-al) | Toward or on the surface of the body. | The ribs are superficial to the lungs. |
| **Deep** (DĒP) | Away from the surface of the body. | The ribs are deep to the skin of the chest. |

[*]In four-legged animals anterior = cephalic (toward the head), ventral = inferior, posterior = caudal (toward the tail), and dorsal = superior.

[†]The midline is an imaginary vertical line that divides the body into equal right and left sides.

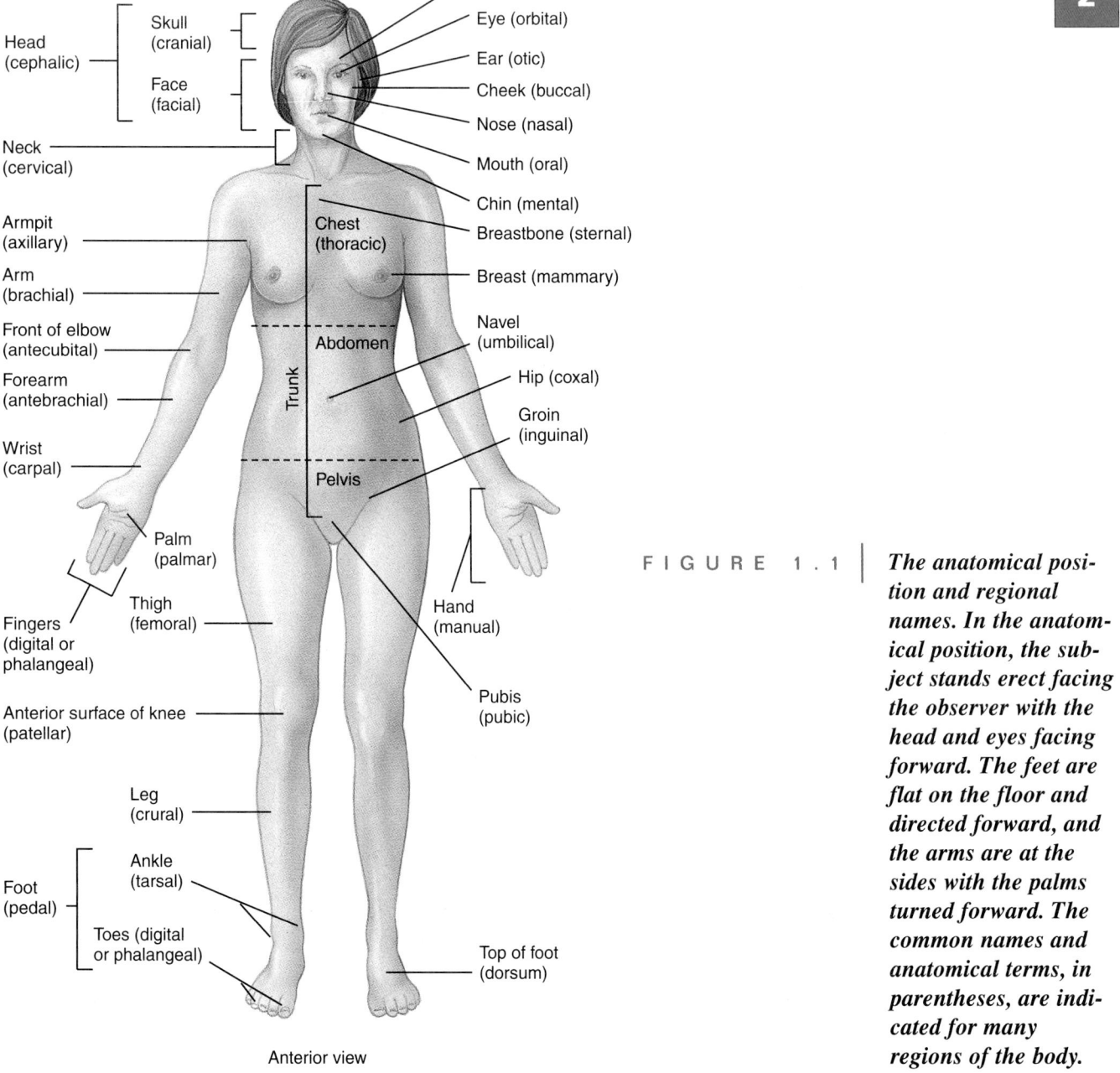

Head (cephalic)

Skull (cranial)

Face (facial)

Neck (cervical)

Armpit (axillary)

Arm (brachial)

Front of elbow (antecubital)

Forearm (antebrachial)

Wrist (carpal)

Palm (palmar)

Fingers (digital or phalangeal)

Thigh (femoral)

Anterior surface of knee (patellar)

Leg (crural)

Foot (pedal)

Ankle (tarsal)

Toes (digital or phalangeal)

Forehead (frontal)

Eye (orbital)

Ear (otic)

Cheek (buccal)

Nose (nasal)

Mouth (oral)

Chin (mental)

Breastbone (sternal)

Chest (thoracic)

Breast (mammary)

Abdomen

Navel (umbilical)

Hip (coxal)

Groin (inguinal)

Trunk

Pelvis

Hand (manual)

Pubis (pubic)

Top of foot (dorsum)

Anterior view

FIGURE 1.1 | *The anatomical position and regional names. In the anatomical position, the subject stands erect facing the observer with the head and eyes facing forward. The feet are flat on the floor and directed forward, and the arms are at the sides with the palms turned forward. The common names and anatomical terms, in parentheses, are indicated for many regions of the body.*

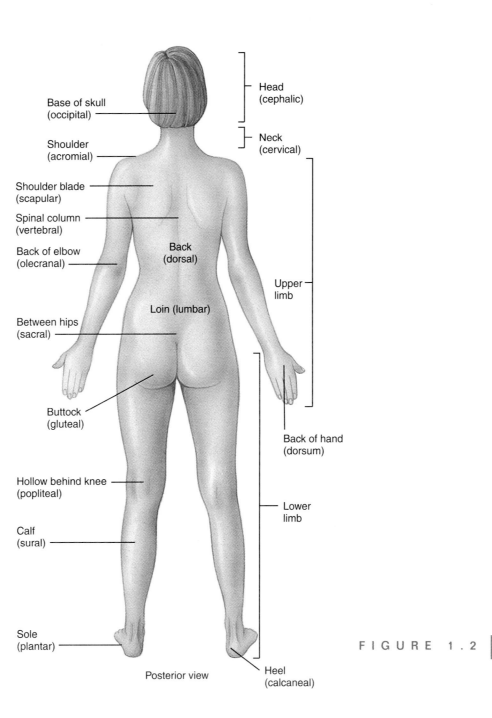

Base of skull (occipital)

Head (cephalic)

Neck (cervical)

Shoulder (acromial)

Shoulder blade (scapular)

Spinal column (vertebral)

Back of elbow (olecranal)

Back (dorsal)

Upper limb

Loin (lumbar)

Between hips (sacral)

Back of hand (dorsum)

Buttock (gluteal)

Hollow behind knee (popliteal)

Lower limb

Calf (sural)

Sole (plantar)

Posterior view

Heel (calcaneal)

FIGURE 1.2 | *The anatomical position and regional names*

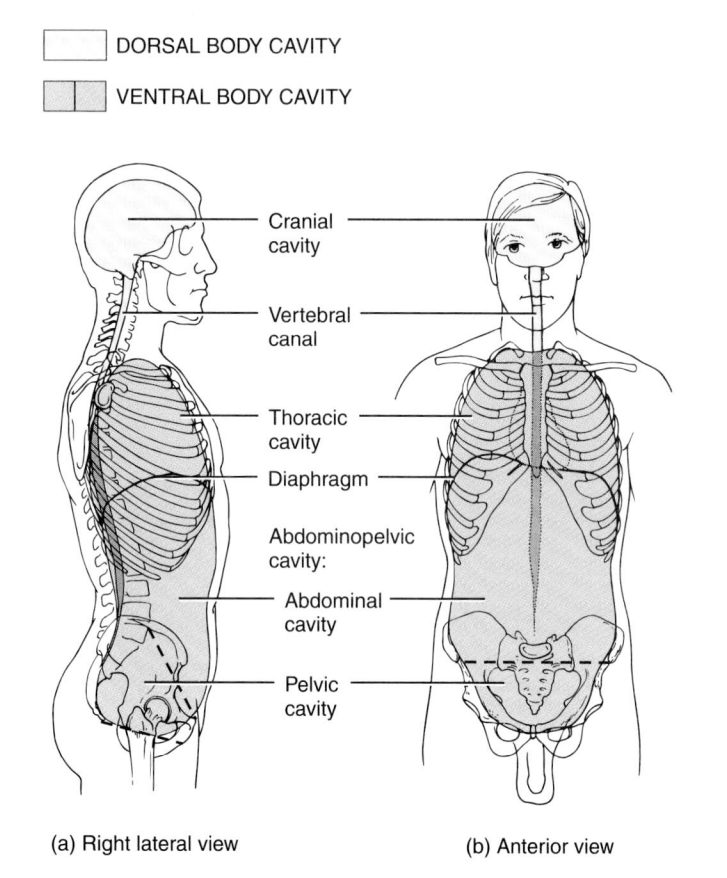

DORSAL BODY CAVITY

VENTRAL BODY CAVITY

Cranial cavity

Vertebral canal

Thoracic cavity

Diaphragm

Abdominopelvic cavity:

Abdominal cavity

Pelvic cavity

(a) Right lateral view

(b) Anterior view

FIGURE 1.3 | *Principal subdivisions of the dorsal and ventral body cavities*

| CAVITY | COMMENTS |
|---|---|
| **Dorsal** | |
| Cranial | Formed by cranial bones and contains brain and its coverings. |
| Vertebral | Formed by vertebral column and contains spinal cord and beginnings of spinal nerves. |
| **Ventral** | |
| Thoracic | Chest cavity; separated from abdominal cavity by diaphragm. |
| Pleural | Contains lungs. |
| Pericardial | Contains heart. |
| Mediastinum | Region between the lungs from the breastbone to backbone that contains heart, thymus gland, esophagus, trachea, bronchi, and many large blood and lymphatic vessels. |
| **Abdominopelvic** | Subdivided into abdominal and pelvic cavities. |
| Abdominal | Contains stomach, spleen, liver, gallbladder, pancreas, small intestine, and most of large intestine. |
| Pelvic | Contains urinary bladder, portions of the large intestine, and internal female and male reproductive organs. |

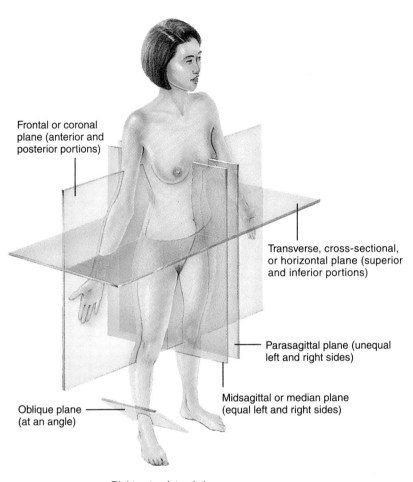

Frontal or coronal plane (anterior and posterior portions)

Transverse, cross-sectional, or horizontal plane (superior and inferior portions)

Parasagittal plane (unequal left and right sides)

Midsagittal or median plane (equal left and right sides)

Oblique plane (at an angle)

Right anterolateral view

FIGURE 1.4 | *Planes are imaginary flat surfaces that divide the entire body or individual organs into various portions. The descriptions in parentheses indicate how each plane divides the body.*

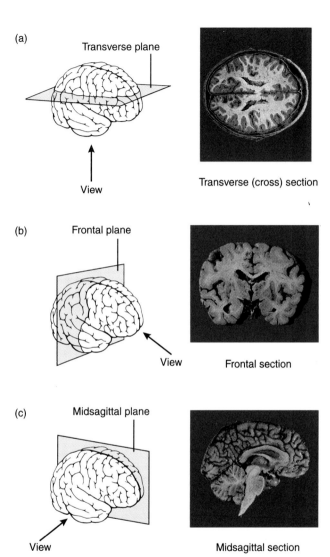

(a) Transverse plane

View

Transverse (cross) section

(b) Frontal plane

View

Frontal section

(c) Midsagittal plane

View

Midsagittal section

FIGURE 1.5 | *Planes and sections. The planes are shown in the diagrams on the left and the sections that result are shown in the photographs of the brain on the right.*

5

**TABLE 2.1 | *Tissues***

| Tissue | Comment | Tissue | Comment |
|---|---|---|---|
| **Epithelial Tissue** | | **Connective Tissue, continued** | |
| I. Covering and lining | Forms outer covering of body and some viscera; lines body cavities, some viscera, blood vessels, and ducts; makes up parts of sense organs. | B. Dense | |
| | | Dense regular | Forms tendons and ligaments. |
| A. Simple | Single layer of cells. | Dense irregular | Found in dermis of skin, fasciae, and membranes around various structures. |
| Squamous | Flat, scalelike cells. | Elastic | Provides stretch and strength. |
| Cuboidal | Cube-shaped cells. | C. Cartilage | Has no blood or nerve supply. |
| Columnar | Rectangular-shaped cells. | Hyaline | Found at ends of long bones and ribs. |
| B. Stratified | Two or more layers of cells. | Fibrocartilage | Found in pubic symphysis and intervertebral discs. |
| Squamous | Flat, scalelike cells in superficial layer. | Elastic | Found in larynx and external ear. |
| Cuboidal | Cube-shaped cells in superficial layer. | D. Bone (osseous) | Contains very rigid intercelluar substance and is classified as compact or spongy. |
| Columnar | Rectangular-shaped cells in superficial layer. | E. Blood (vascular) | Liquid connective tissue consisting of plasma and formed elements (red blood cells, white blood cells, and platelets). |
| Transitional | Cells variable in shape. | | |
| C. Pseudostratified | Single layer of cells that appears to be stratified. | | |
| II. Glandular | Forms secretory portions of glands. | **Muscle Tissue** | Highly specialized for contraction. |
| A. Exocrine | Secrete products into ducts. | I. Skeletal | Usually attached to bones, striated, voluntary. |
| B. Endocrine | Secrete hormones into the blood. | II. Cardiac | Found in the heart, striated, involuntary. |
| | | III. Smooth (visceral) | Found in viscera and blood vessels, nonstriated, involuntary. |
| **Connective Tissue** | | | |
| I. Embryonic | Present primarily in embryo and fetus. | **Nervous Tissue** | |
| A. Mesenchyme | Embryonic tissue from which all other connective tissues develop. | I. Neurons | Specialized for detecting stimuli, converting them into action potentials, and conducting action potentials. |
| B. Mucous | Fetal tissue found in umbilical cord. | II. Neuroglia | Protect and support neurons. |
| II. Mature | Found in newborn. | | |
| A. Loose | | | |
| Areolar | One of the most abundant connective tissues. | | |
| Adipose | Specialized for fat storage. | | |
| Reticular | Forms stroma (framework) of certain organs. | | |

FIGURE 2.1 | *Histology of epithelial tissues*

Plasma membrane

Cytoplasm

Nucleus

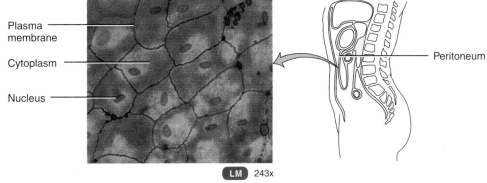

LM 243x

Surface view of mesothelial lining of peritoneum

Peritoneum

Simple squamous cell

Basement membrane

Connective tissue

(a) Simple squamous epithelium

Kidney

Simple cuboidal epithelium

Lumen of tubule

Nucleus of simple cuboidal cell

Simple cuboidal cell

Basement membrane

Connective tissue

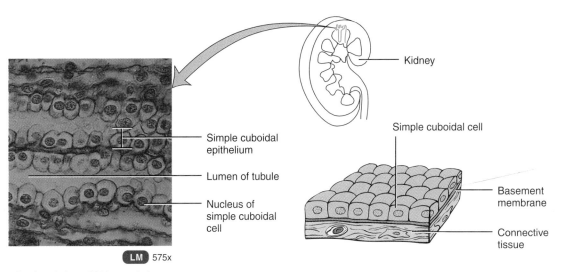

LM 575x

Sectional view of kidney tubules

(b) Simple cuboidal epithelium

7

FIGURE 2.1 *Histology of epithelial tissues, continued*

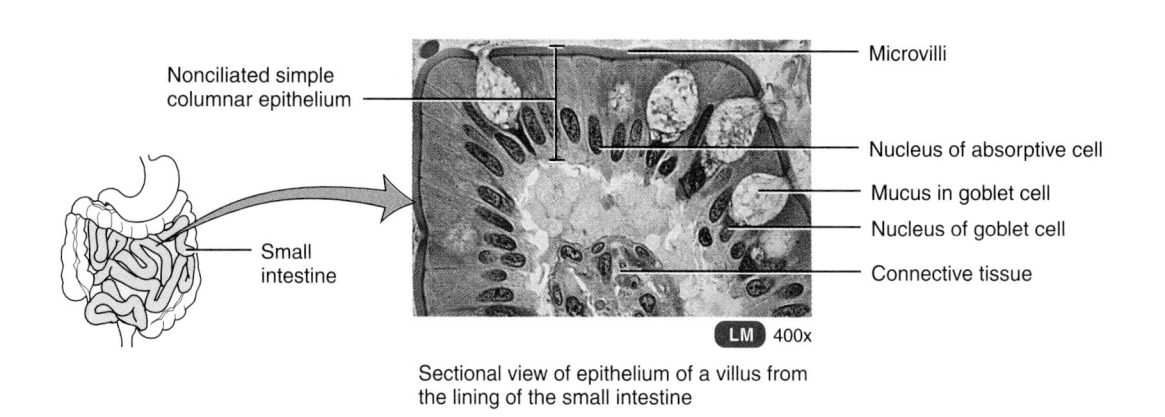

Nonciliated simple columnar epithelium

Small intestine

Microvilli

Nucleus of absorptive cell

Mucus in goblet cell

Nucleus of goblet cell

Connective tissue

LM 400x

Sectional view of epithelium of a villus from the lining of the small intestine

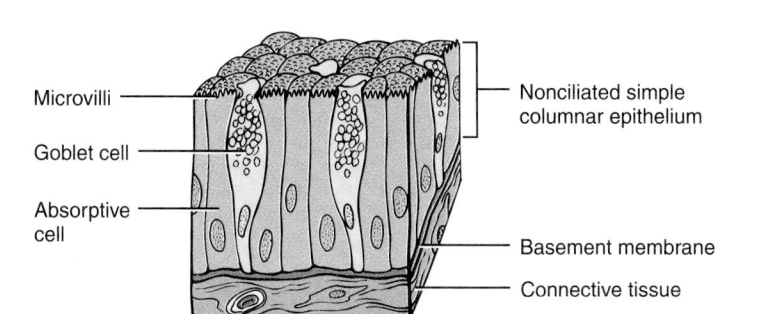

Microvilli

Goblet cell

Absorptive cell

Nonciliated simple columnar epithelium

Basement membrane

Connective tissue

(c) Nonciliated simple columnar epithelium

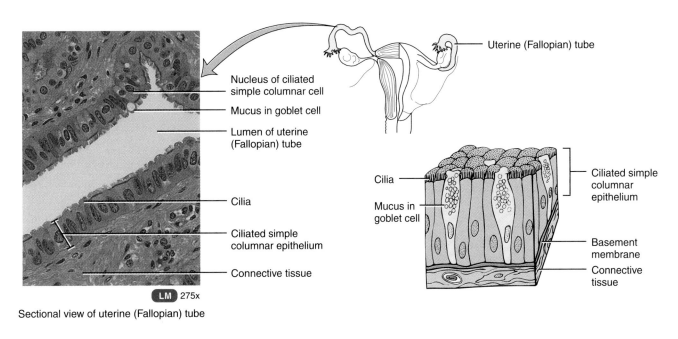

Nucleus of ciliated
simple columnar cell

Mucus in goblet cell

Lumen of uterine
(Fallopian) tube

Cilia

Ciliated simple
columnar epithelium

Connective tissue

LM 275x

Sectional view of uterine (Fallopian) tube

Uterine (Fallopian) tube

Cilia

Mucus in
goblet cell

Ciliated simple
columnar
epithelium

Basement
membrane

Connective
tissue

(d) Ciliated simple columnar epithelium

FIGURE 2.1 *Histology of epithelial tissues, continued*

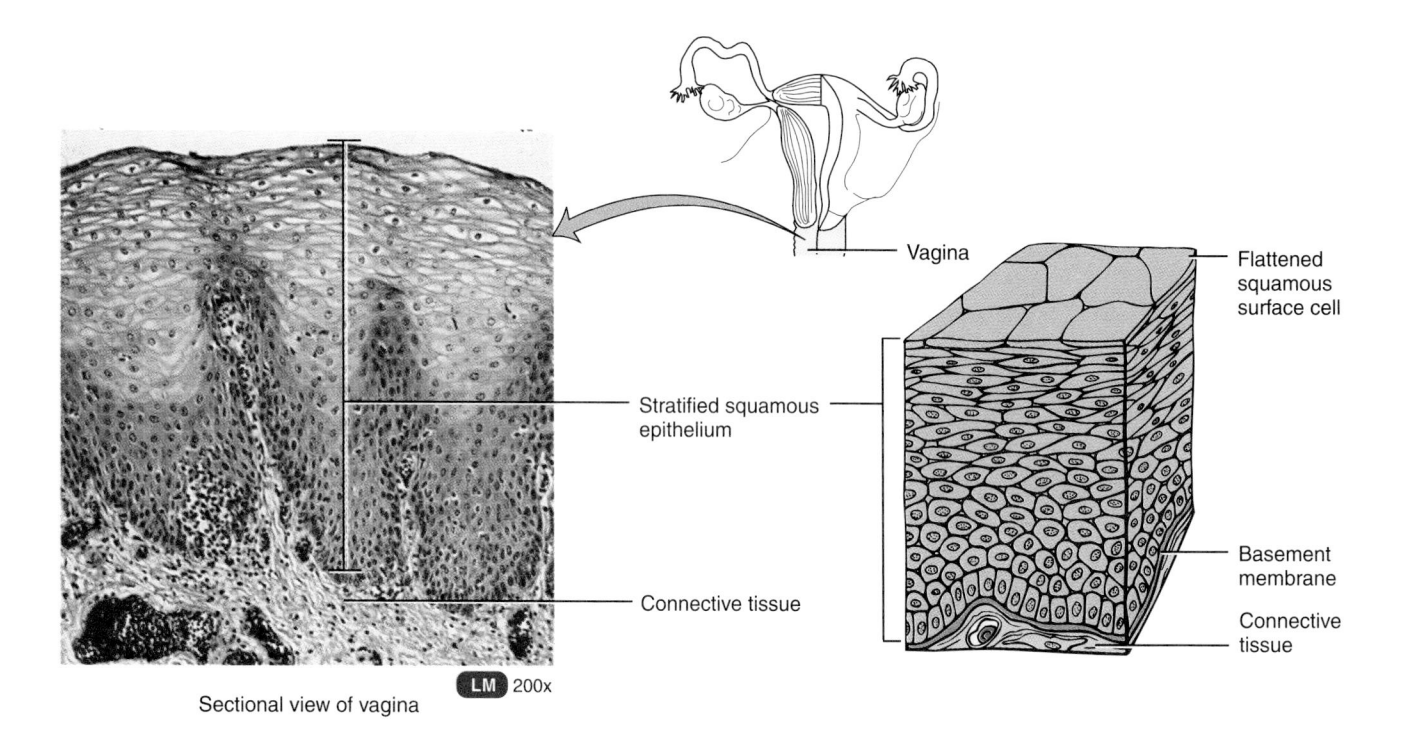

Vagina

Stratified squamous epithelium

Connective tissue

Flattened squamous surface cell

Basement membrane

Connective tissue

LM 200x

Sectional view of vagina

(e) Stratified squamous epithelium

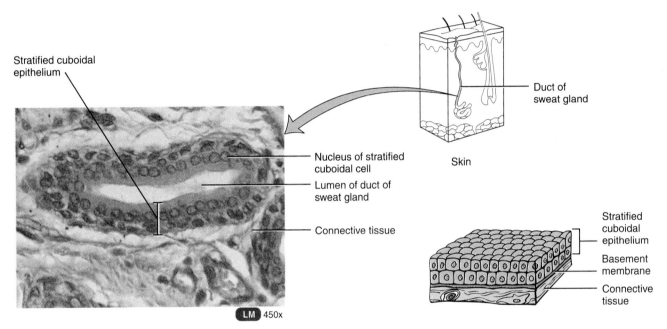

Stratified cuboidal
epithelium

Nucleus of stratified
cuboidal cell

Lumen of duct of
sweat gland

Connective tissue

Duct of
sweat gland

Skin

Stratified
cuboidal
epithelium

Basement
membrane

Connective
tissue

LM 450x

Sectional view of the duct of a sweat gland

(f) Stratified cuboidal epithelium

11

F I G U R E   2 . 1    *Histology of epithelial tissues, continued*

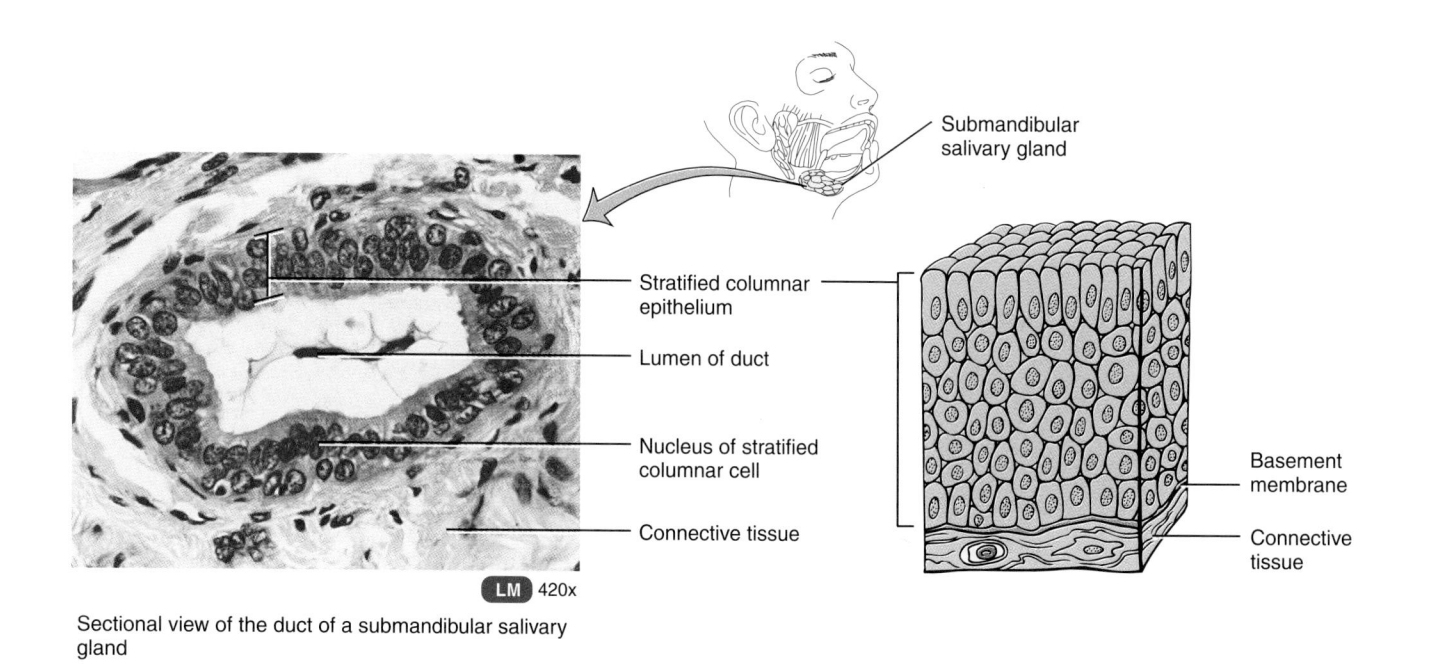

Submandibular salivary gland

Stratified columnar epithelium

Lumen of duct

Nucleus of stratified columnar cell

Connective tissue

Basement membrane

Connective tissue

LM 420x

Sectional view of the duct of a submandibular salivary gland

(g) Stratified columnar epithelium

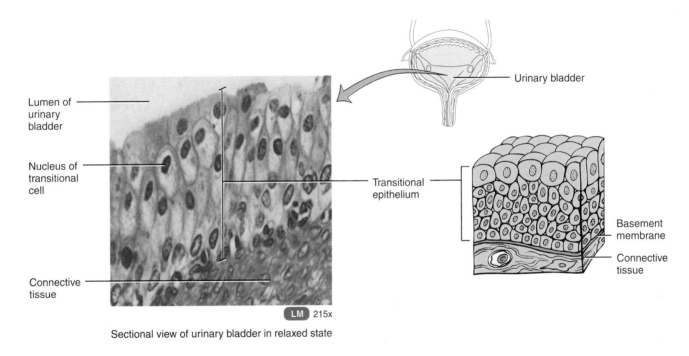

Lumen of
urinary
bladder

Nucleus of
transitional
cell

Connective
tissue

Urinary bladder

Transitional
epithelium

Basement
membrane

Connective
tissue

LM 215x

Sectional view of urinary bladder in relaxed state

(h) Transitional epithelium (relaxed)

13

FIGURE 2.1 *Histology of epithelial tissues, continued*

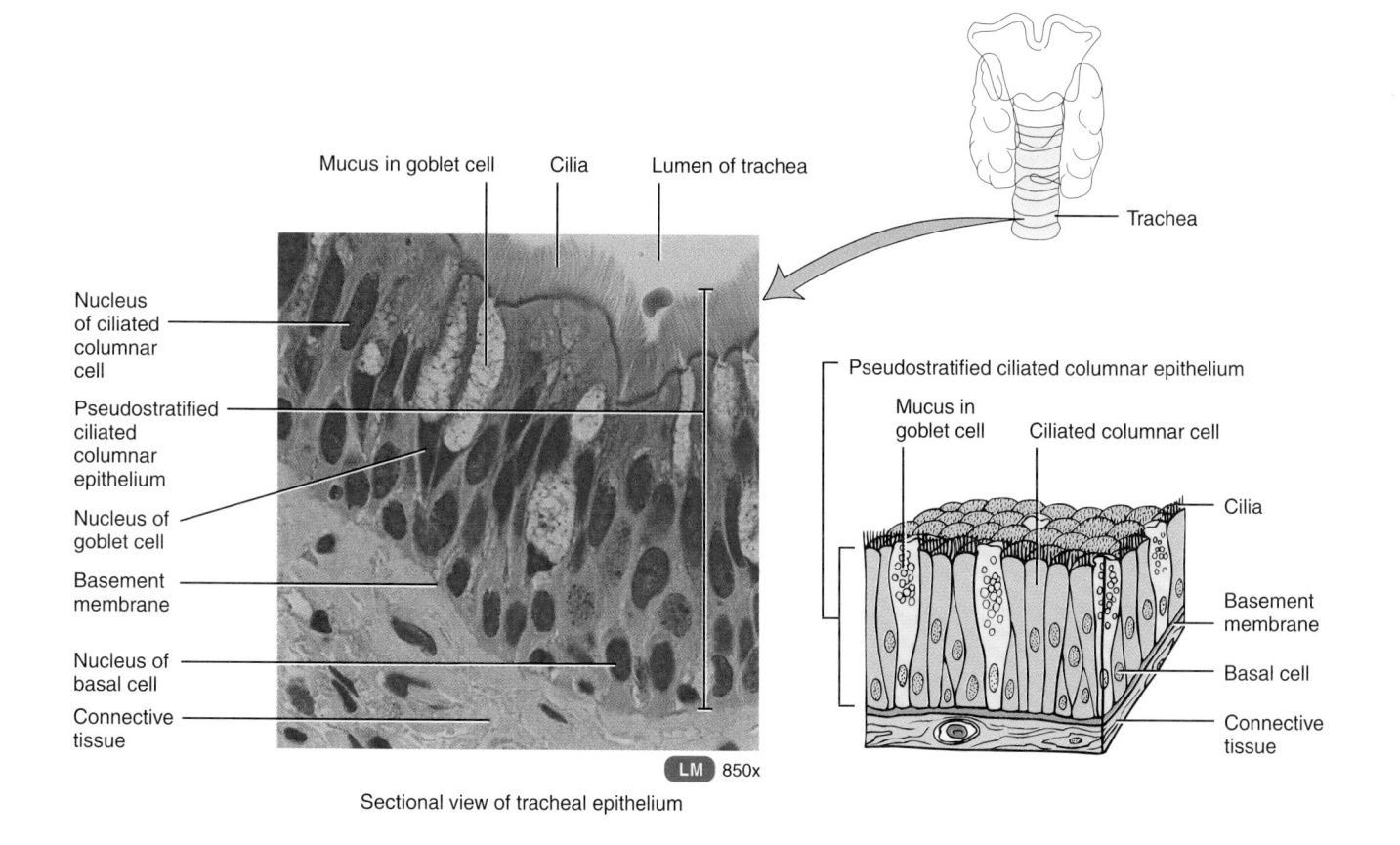

Mucus in goblet cell

Cilia

Lumen of trachea

Trachea

Nucleus of ciliated columnar cell

Pseudostratified ciliated columnar epithelium

Nucleus of goblet cell

Basement membrane

Nucleus of basal cell

Connective tissue

LM 850x

Sectional view of tracheal epithelium

Pseudostratified ciliated columnar epithelium

Mucus in goblet cell

Ciliated columnar cell

Cilia

Basement membrane

Basal cell

Connective tissue

(i) Pseudostratified ciliated columnar epithelium

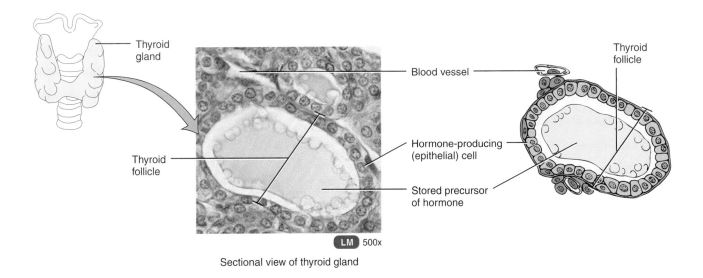

Thyroid gland

Blood vessel

Thyroid follicle

Thyroid follicle

Hormone-producing (epithelial) cell

Stored precursor of hormone

LM 500x

Sectional view of thyroid gland

(j) Endocrine gland: glandular epithelium

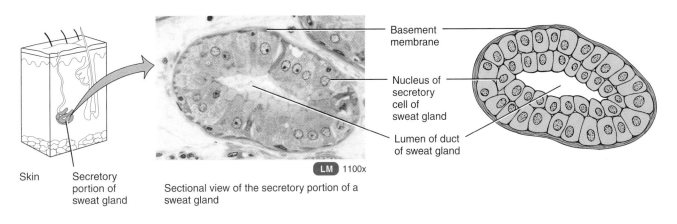

Basement membrane

Nucleus of secretory cell of sweat gland

Lumen of duct of sweat gland

LM 1100x

Skin

Secretory portion of sweat gland

Sectional view of the secretory portion of a sweat gland

(k) Exocrine gland: glandular epithelium

15

FIGURE 2.2 **Histology of connective tissue**

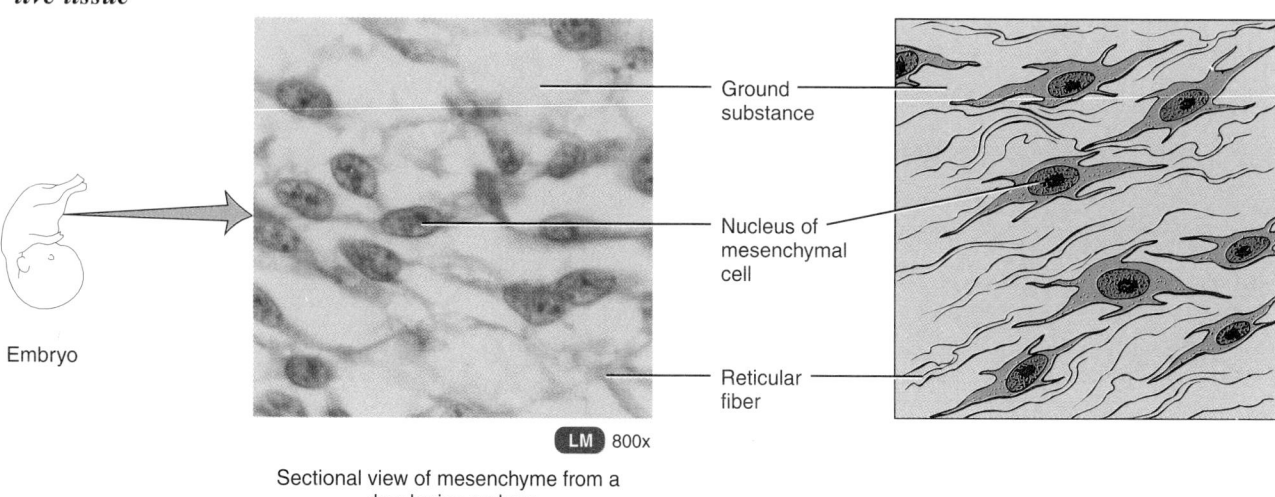

Embryo

Ground substance

Nucleus of mesenchymal cell

Reticular fiber

LM 800x

Sectional view of mesenchyme from a developing embryo

(a) Mesenchyme

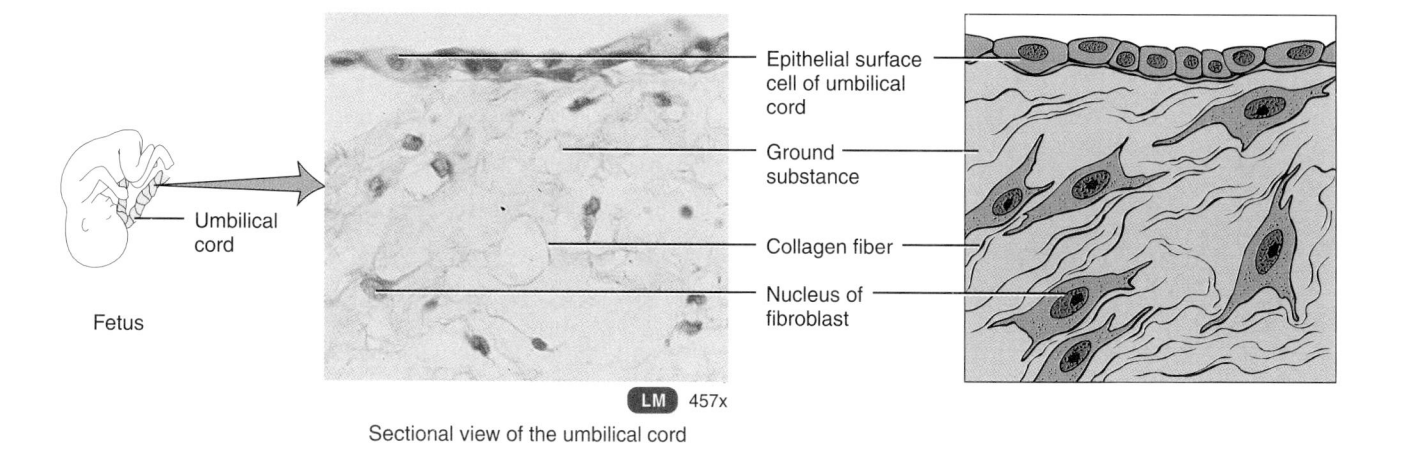

Fetus

Umbilical cord

Epithelial surface cell of umbilical cord

Ground substance

Collagen fiber

Nucleus of fibroblast

LM 457x

Sectional view of the umbilical cord

(b) Mucous connective tissue

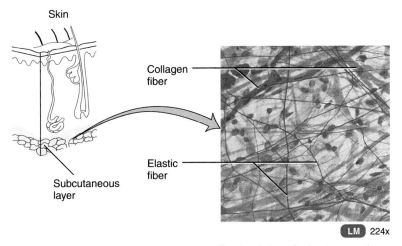

Skin

Collagen
fiber

Elastic
fiber

Subcutaneous
layer

Macrophage

Reticular fiber

Fibroblast

Elastic fiber

Collagen fiber

Mast cell

Plasma cell

LM 224x

Sectional view of subcutaneous tissue

(c) Areolar connective tissue

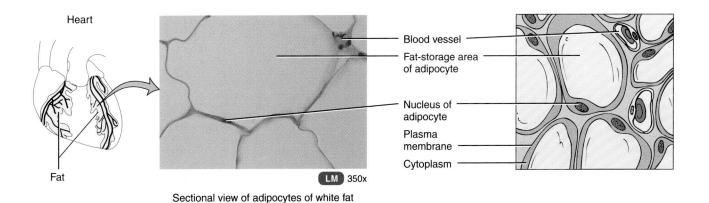

Heart

Fat

Blood vessel

Fat-storage area
of adipocyte

Nucleus of
adipocyte

Plasma
membrane

Cytoplasm

LM 350x

Sectional view of adipocytes of white fat

(d) Adipose tissue

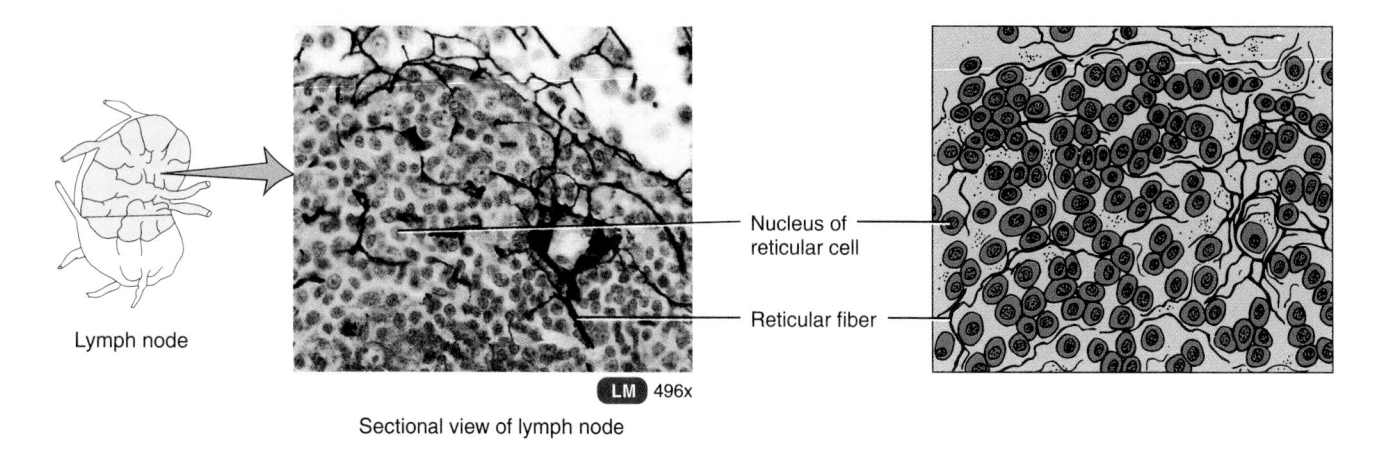

Lymph node

Sectional view of lymph node

Nucleus of reticular cell

Reticular fiber

LM 496x

(e) Reticular connective tissue

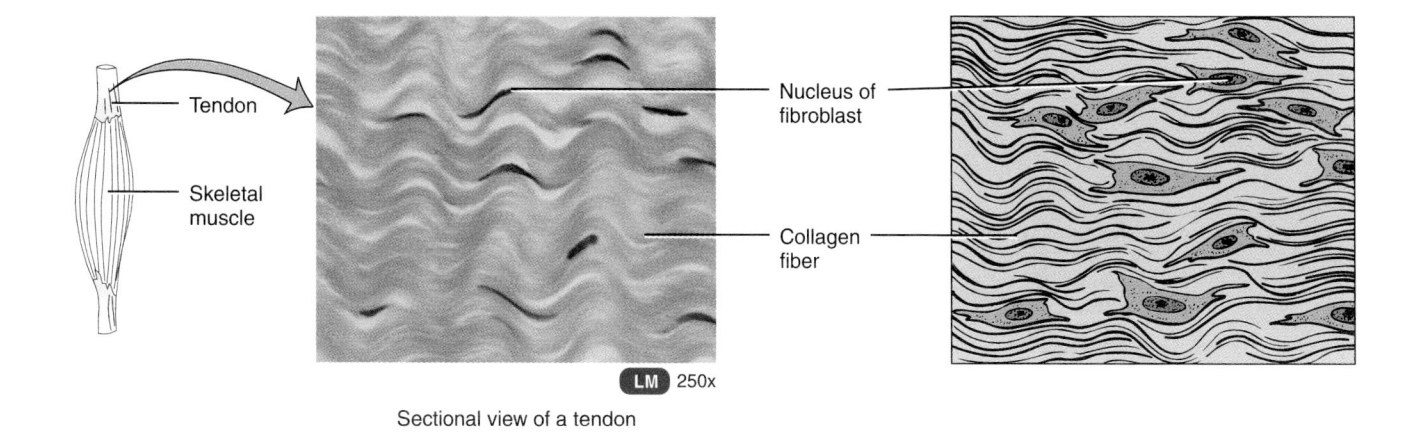

Tendon

Skeletal muscle

Nucleus of fibroblast

Collagen fiber

LM 250x

Sectional view of a tendon

(f) Dense regular connective tissue

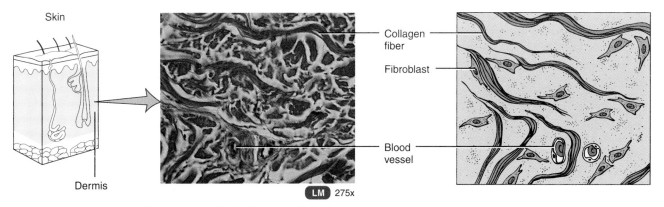

Skin

Dermis

Collagen fiber

Fibroblast

Blood vessel

LM 275x

Sectional view of reticular region of dermis of skin

(g) Dense irregular connective tissue

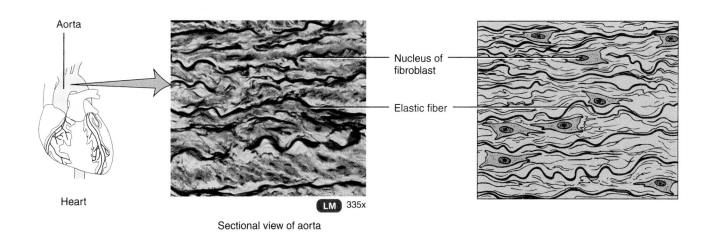

Aorta

Heart

Nucleus of fibroblast

Elastic fiber

LM 335x

Sectional view of aorta

(h) Elastic connective tissue

19

FIGURE 2 : 2 *Histology of connective tissue, continued*

20

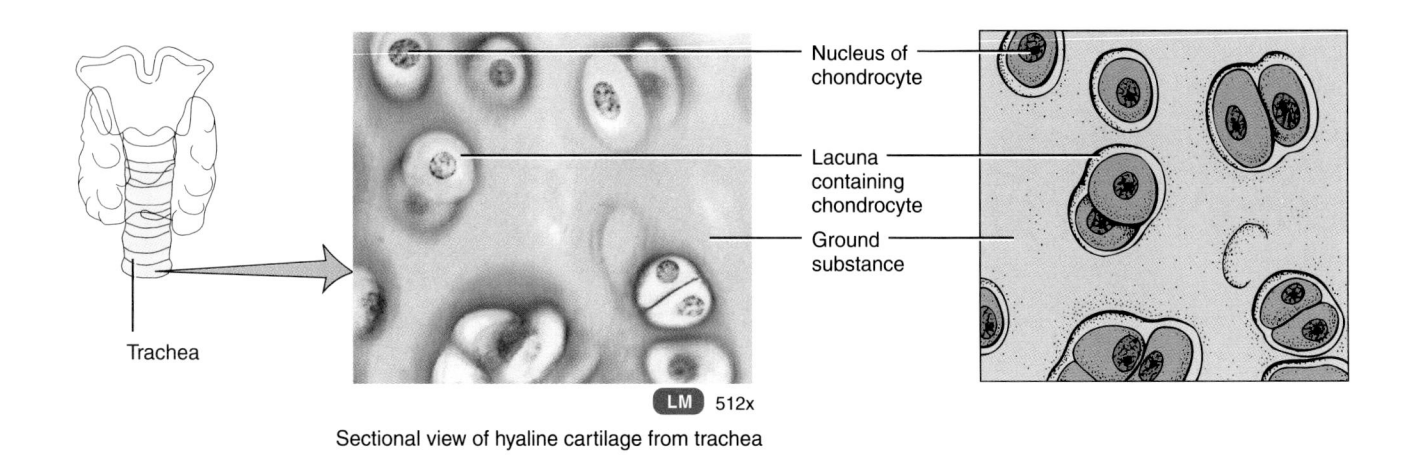

Nucleus of
chondrocyte

Lacuna
containing
chondrocyte

Ground
substance

Trachea

LM 512x

Sectional view of hyaline cartilage from trachea

(i) Hyaline cartilage

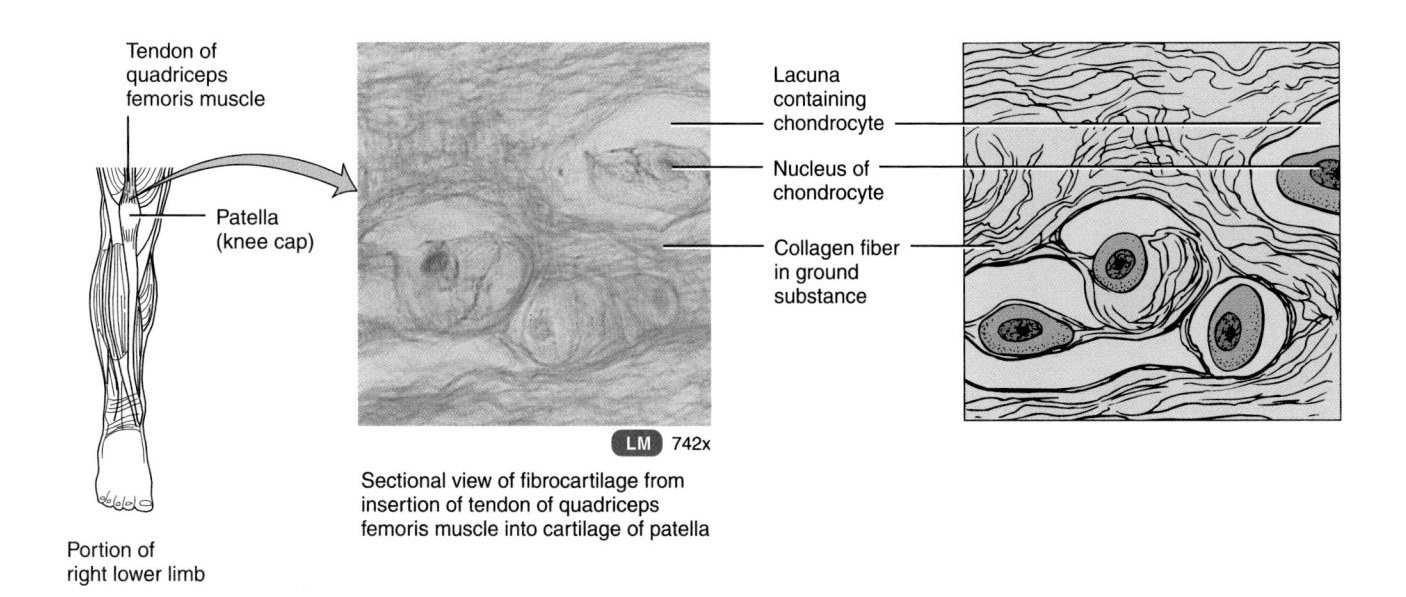

Tendon of
quadriceps
femoris muscle

Patella
(knee cap)

Lacuna
containing
chondrocyte

Nucleus of
chondrocyte

Collagen fiber
in ground
substance

LM 742x

Sectional view of fibrocartilage from
insertion of tendon of quadriceps
femoris muscle into cartilage of patella

Portion of
right lower limb

(j) Fibrocartilage

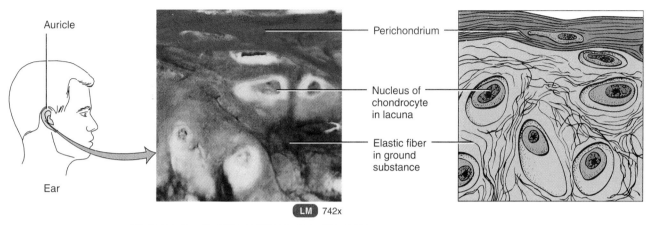

Auricle

Ear

Perichondrium

Nucleus of
chondrocyte
in lacuna

Elastic fiber
in ground
substance

LM 742x

Sectional view of elastic cartilage from auricle of ear

(k) Elastic cartilage

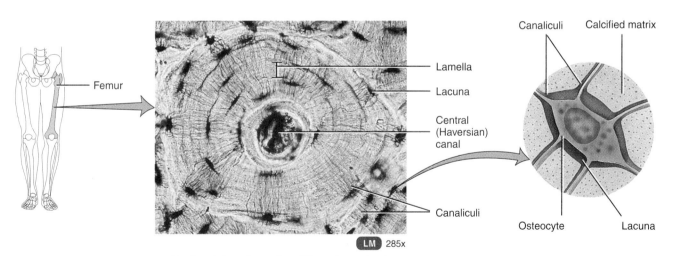

Femur

Lamella

Lacuna

Central
(Haversian)
canal

Canaliculi

Canaliculi    Calcified matrix

Osteocyte        Lacuna

LM 285x

Sectional view of an osteon (Haversian system)
from the femur (thigh bone)

(l) Bone tissue

*Histology of connective tissue, continued*

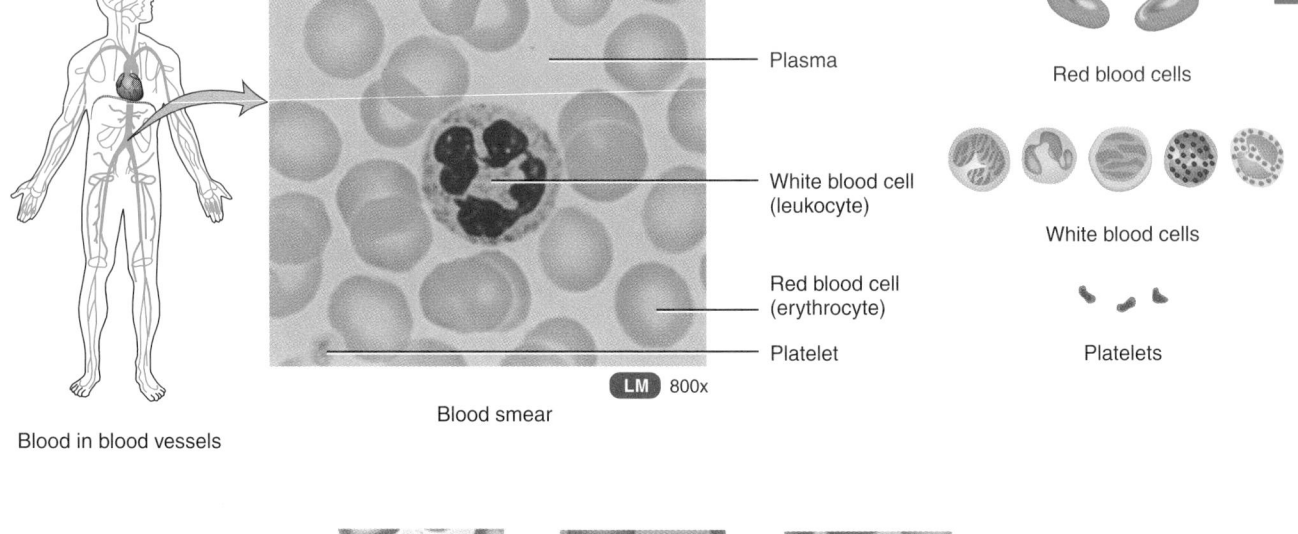

Plasma

White blood cell (leukocyte)

Red blood cell (erythrocyte)

Platelet

Red blood cells

White blood cells

Platelets

LM  800x

Blood smear

Blood in blood vessels

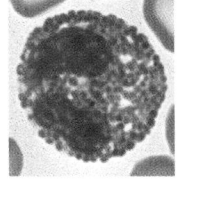

Eosinophil

Basophil

Neutrophil

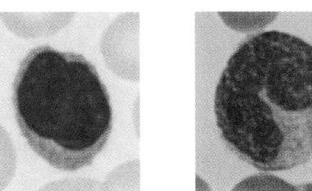

Large lymphocyte

Monocyte

LM  all at 1500x

Details of a blood smear

(m) Blood tissue

FIGURE 2.3 *Histology of muscle tissue*

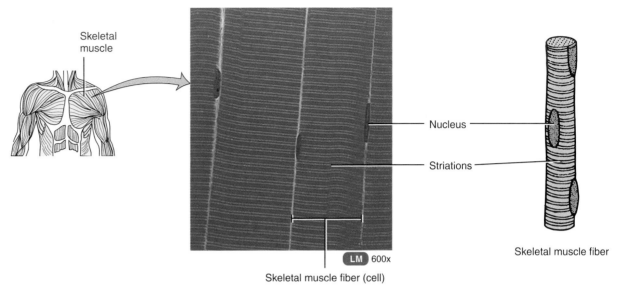

Skeletal muscle

Nucleus

Striations

LM 600x

Skeletal muscle fiber (cell)

Skeletal muscle fiber

(a) Longitudinal section of skeletal muscle tissue

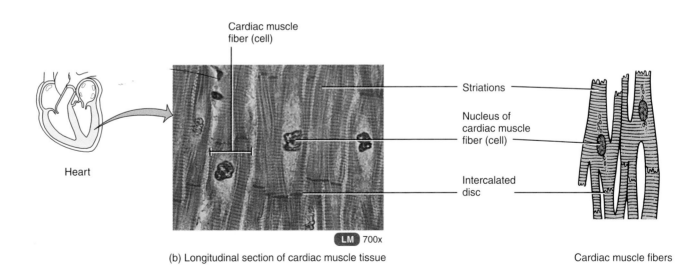

Cardiac muscle fiber (cell)

Striations

Nucleus of cardiac muscle fiber (cell)

Intercalated disc

Heart

LM 700x

(b) Longitudinal section of cardiac muscle tissue

Cardiac muscle fibers

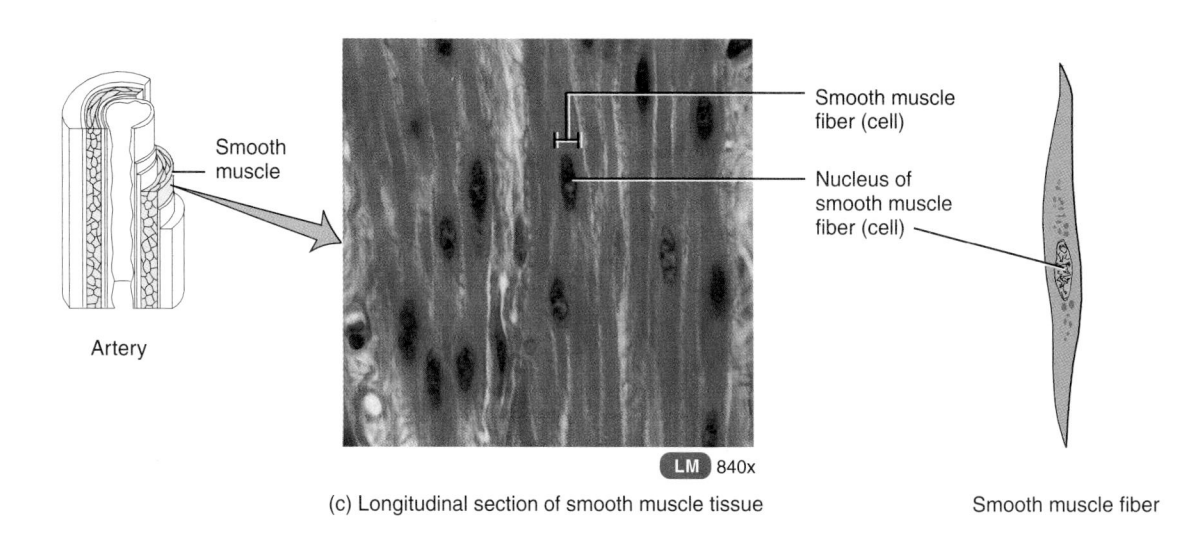

Smooth muscle

Artery

Smooth muscle fiber (cell)

Nucleus of smooth muscle fiber (cell)

LM 840x

(c) Longitudinal section of smooth muscle tissue

Smooth muscle fiber

FIGURE 2.3 *Histology of muscle tissue, continued*

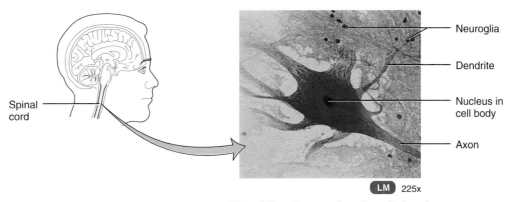

Spinal cord

Neuroglia

Dendrite

Nucleus in cell body

Axon

LM 225x

Motor (efferent) neuron from the spinal cord and neuroglia

FIGURE 2.4 | *Histology of nervous tissue and neuroglia*

**TABLE 3.1** | *Divisions of the Adult Skeletal System*

| REGIONS OF THE SKELETON | NUMBER OF BONES | REGIONS OF THE SKELETON | NUMBER OF BONES |
|---|---|---|---|
| **Axial Skeleton** | | **Appendicular Skeleton** | |
| Skull | | Pectoral (shoulder) girdles | |
|   *Cranium* | 8 |   *Clavicle* | 2 |
|   *Face* | 14 |   *Scapula* | 2 |
| Hyoid | 1 | Upper limbs (extremities) | |
| Auditory ossicles | 6 |   *Humerus* | 2 |
| Vertebral column | 26 |   *Ulna* | 2 |
| Thorax | |   *Radius* | 2 |
|   *Sternum* | 1 |   *Carpals* | 16 |
|   *Ribs* | 24 |   *Metacarpals* | 10 |
| | Subtotal = 80 |   *Phalanges* | 28 |
| | | Pelvic (hip) girdle | |
| | |   *Hip, pelvic, or coxal bone* | 2 |
| | | Lower limbs (extremities) | |
| | |   *Femur* | 2 |
| | |   *Fibula* | 2 |
| | |   *Tibia* | 2 |
| | |   *Patella* | 2 |
| | |   *Tarsals* | 14 |
| | |   *Metatarsals* | 10 |
| | |   *Phalanges* | 28 |
| | | Subtotal = | 126 |
| | | Total = | 206 |

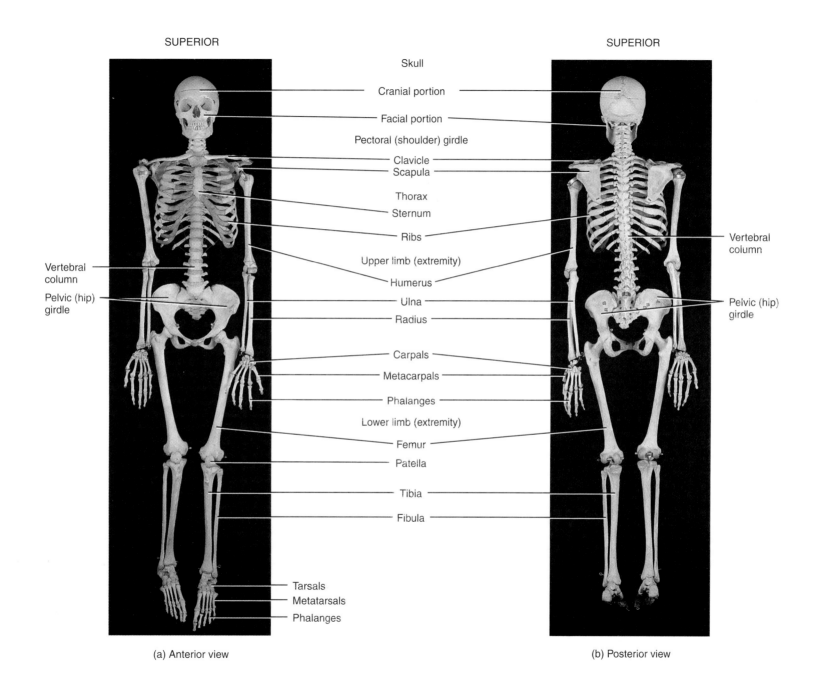

SUPERIOR

SUPERIOR

Skull

Cranial portion

Facial portion

Pectoral (shoulder) girdle

Clavicle

Scapula

Thorax

Sternum

Ribs

Vertebral column

Upper limb (extremity)

Vertebral column

Humerus

Pelvic (hip) girdle

Ulna

Radius

Pelvic (hip) girdle

Carpals

Metacarpals

Phalanges

Lower limb (extremity)

Femur

Patella

Tibia

Fibula

Tarsals

Metatarsals

Phalanges

(a) Anterior view

(b) Posterior view

FIGURE  3 . 1    *Complete skeleton*

27

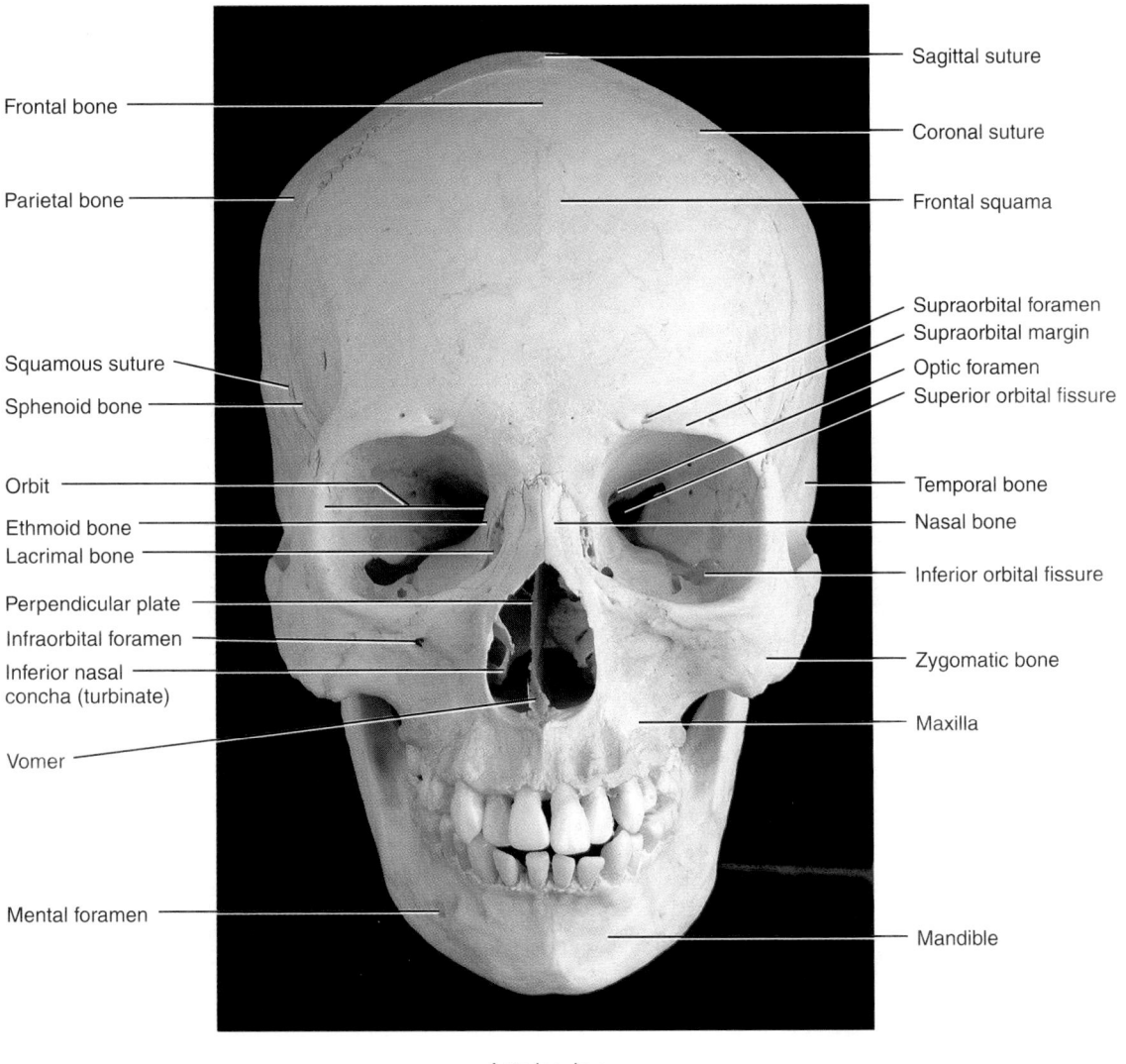

Sagittal suture

Frontal bone

Coronal suture

Parietal bone

Frontal squama

Supraorbital foramen
Supraorbital margin

Squamous suture

Optic foramen
Superior orbital fissure

Sphenoid bone

Orbit

Temporal bone

Ethmoid bone

Nasal bone

Lacrimal bone

Inferior orbital fissure

Perpendicular plate

Infraorbital foramen

Zygomatic bone

Inferior nasal
concha (turbinate)

Vomer

Maxilla

Mental foramen

Mandible

Anterior view

FIGURE 3.2 | *Skull*

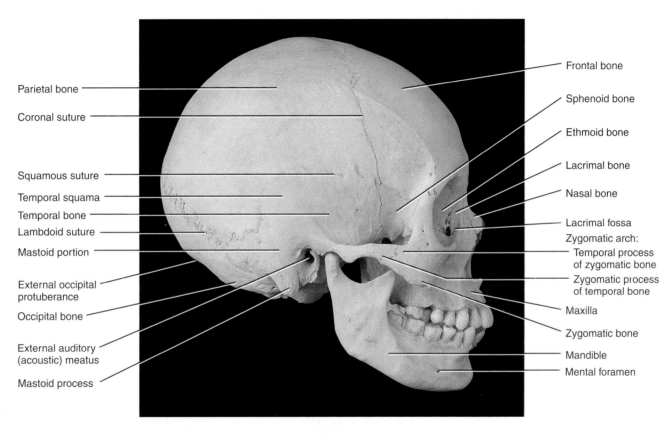

Parietal bone

Coronal suture

Squamous suture

Temporal squama

Temporal bone

Lambdoid suture

Mastoid portion

External occipital
protuberance

Occipital bone

External auditory
(acoustic) meatus

Mastoid process

Frontal bone

Sphenoid bone

Ethmoid bone

Lacrimal bone

Nasal bone

Lacrimal fossa

Zygomatic arch:
  Temporal process
  of zygomatic bone

  Zygomatic process
  of temporal bone

Maxilla

Zygomatic bone

Mandible

Mental foramen

Right lateral view

FIGURE 3.3 | *Skull*

29

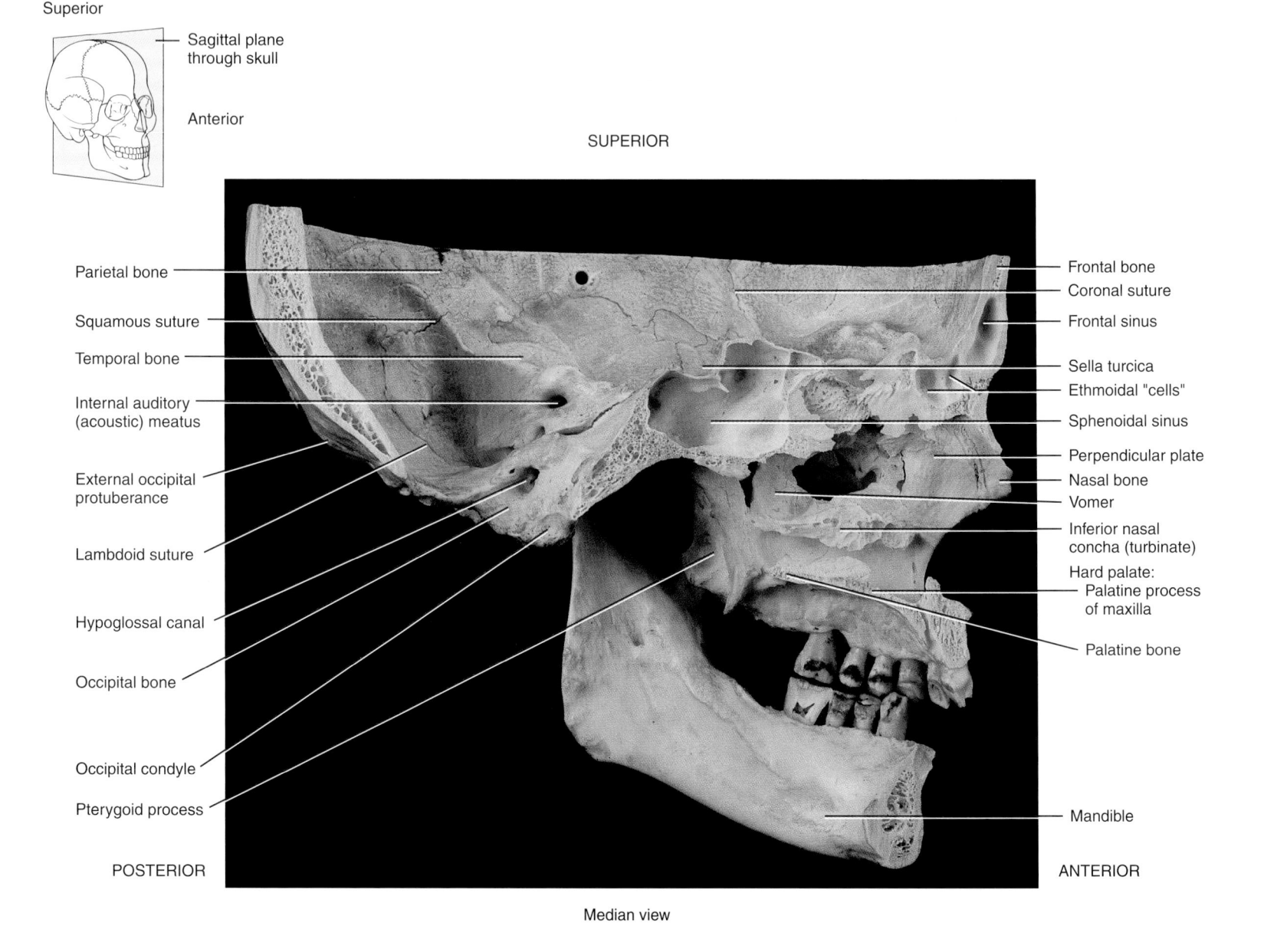

Superior

Sagittal plane
through skull

Anterior

SUPERIOR

Parietal bone

Squamous suture

Temporal bone

Internal auditory
(acoustic) meatus

External occipital
protuberance

Lambdoid suture

Hypoglossal canal

Occipital bone

Occipital condyle

Pterygoid process

POSTERIOR

Frontal bone

Coronal suture

Frontal sinus

Sella turcica

Ethmoidal "cells"

Sphenoidal sinus

Perpendicular plate

Nasal bone

Vomer

Inferior nasal
concha (turbinate)

Hard palate:
Palatine process
of maxilla

Palatine bone

Mandible

ANTERIOR

Median view

FIGURE  3.4  | *Skull*

SUPERIOR

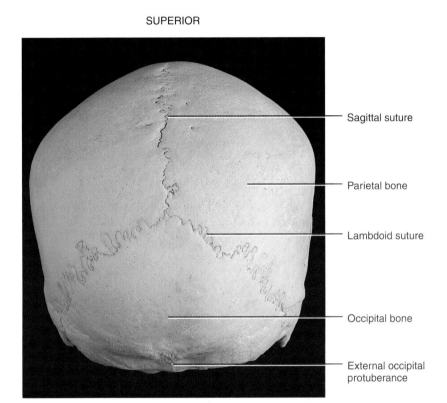

Sagittal suture

Parietal bone

Lambdoid suture

Occipital bone

External occipital
protuberance

INFERIOR

Posterior view

FIGURE 3.5 | *Skull*

ANTERIOR

Frontal bone

Coronal suture

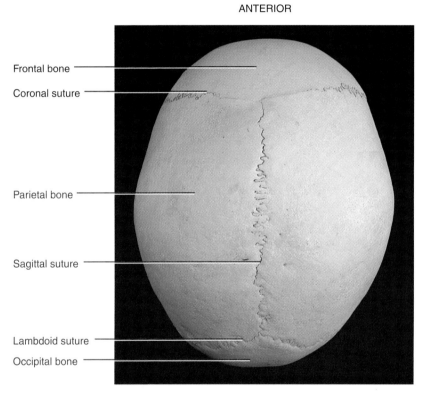

Parietal bone

Sagittal suture

Lambdoid suture

Occipital bone

POSTERIOR

Superior view

FIGURE 3.6 | *Skull*

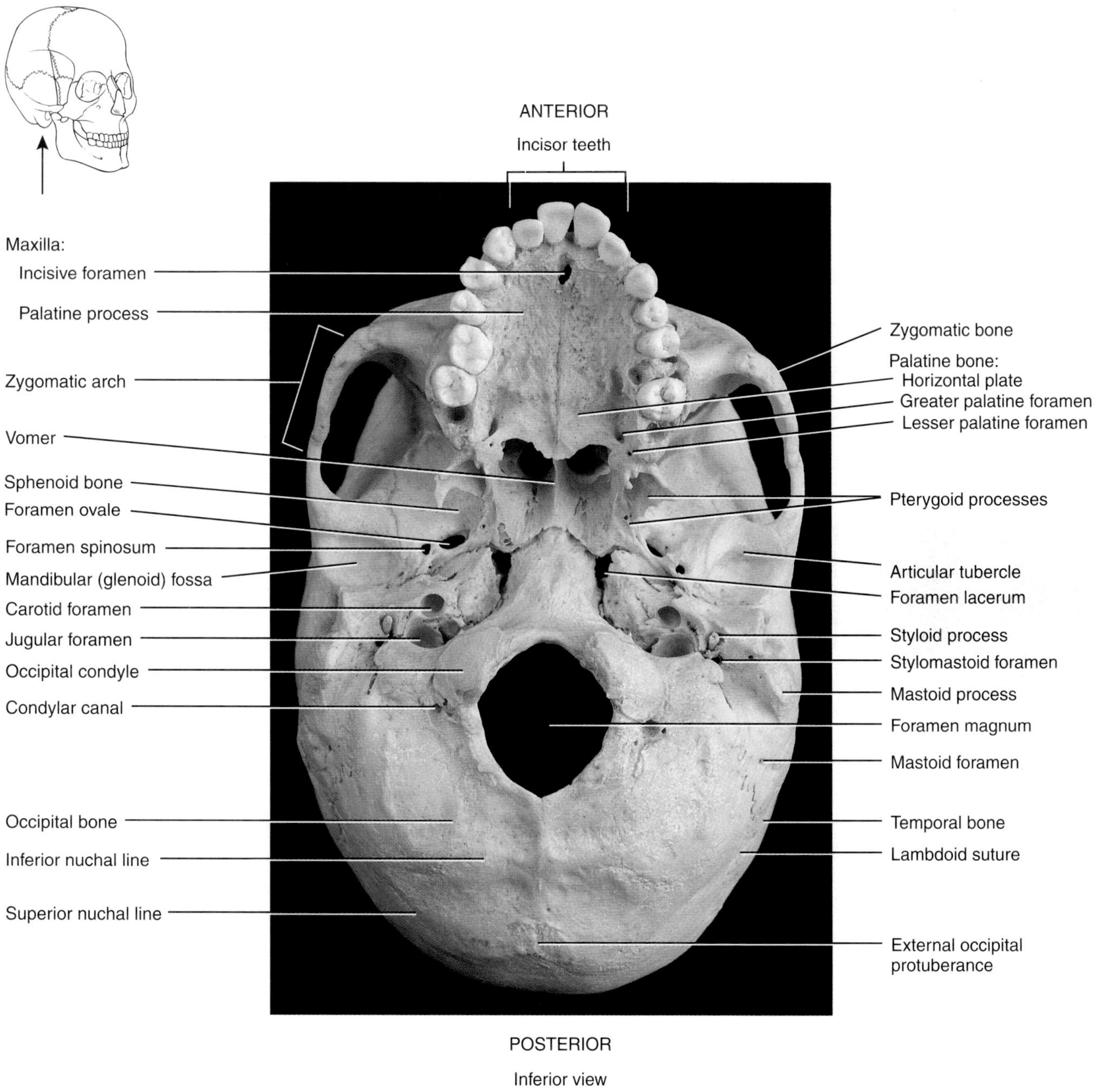

32

ANTERIOR

Incisor teeth

Maxilla:
Incisive foramen
Palatine process

Zygomatic arch

Vomer

Sphenoid bone
Foramen ovale

Foramen spinosum
Mandibular (glenoid) fossa
Carotid foramen
Jugular foramen
Occipital condyle
Condylar canal

Occipital bone
Inferior nuchal line

Superior nuchal line

Zygomatic bone
Palatine bone:
Horizontal plate
Greater palatine foramen
Lesser palatine foramen

Pterygoid processes

Articular tubercle
Foramen lacerum
Styloid process
Stylomastoid foramen
Mastoid process
Foramen magnum
Mastoid foramen

Temporal bone
Lambdoid suture

External occipital
protuberance

POSTERIOR

Inferior view

FIGURE   3 . 7   | *Skull*

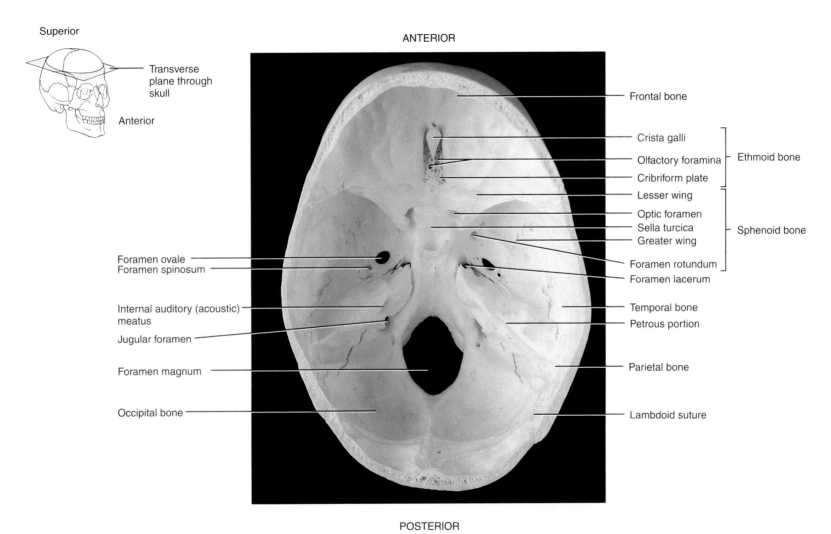

Superior

Transverse
plane through
skull

Anterior

ANTERIOR

Frontal bone

Crista galli

Olfactory foramina          Ethmoid bone

Cribriform plate

Lesser wing

Optic foramen

Sella turcica              Sphenoid bone

Greater wing

Foramen ovale                              Foramen rotundum
Foramen spinosum                           Foramen lacerum

Internal auditory (acoustic)              Temporal bone
meatus                                     Petrous portion

Jugular foramen

Foramen magnum                             Parietal bone

Occipital bone                             Lambdoid suture

POSTERIOR

Superior view of floor of cranium

FIGURE 3.8 | *Skull*

FIGURE 3.9 | *Sphenoid bone*

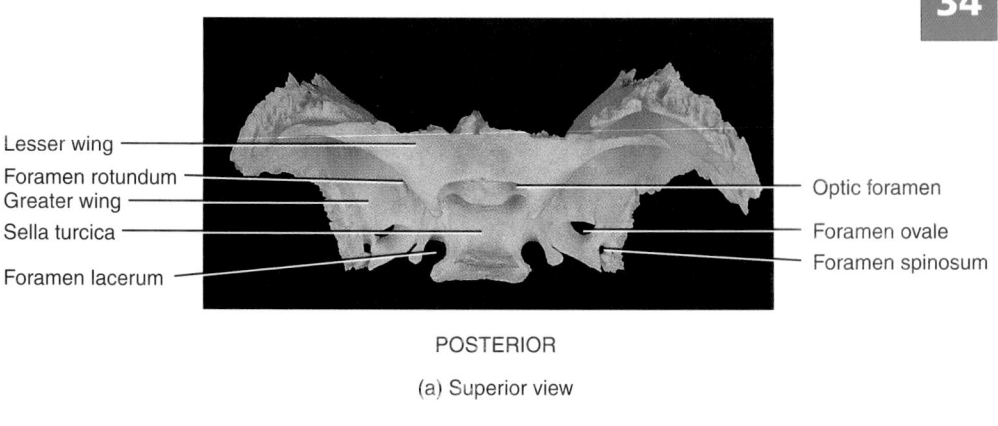

34

ANTERIOR

Lesser wing
Foramen rotundum
Greater wing
Sella turcica
Foramen lacerum
Optic foramen
Foramen ovale
Foramen spinosum

POSTERIOR

(a) Superior view

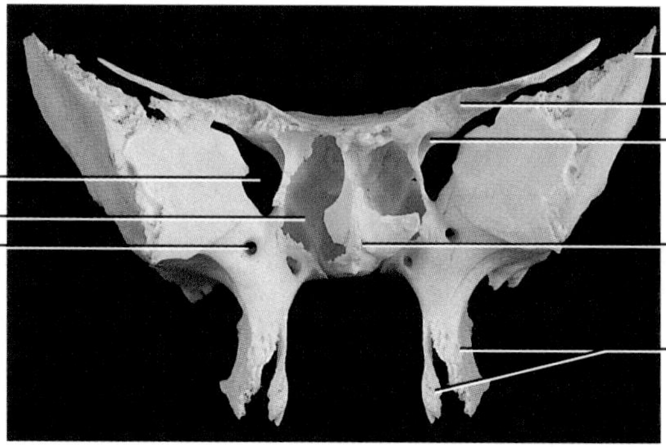

Superior

Frontal plane through skull

Anterior

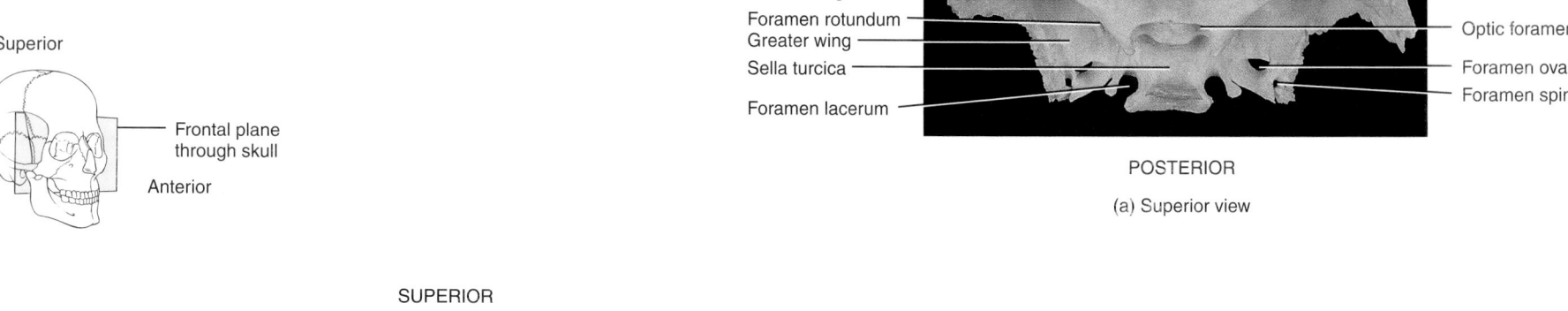

SUPERIOR

Greater wing
Lesser wing
Optic foramen

Superior orbital fissure
Sphenoidal sinus
Foramen rotundum
Body

Pterygoid processes

INFERIOR

(b) Anterior view

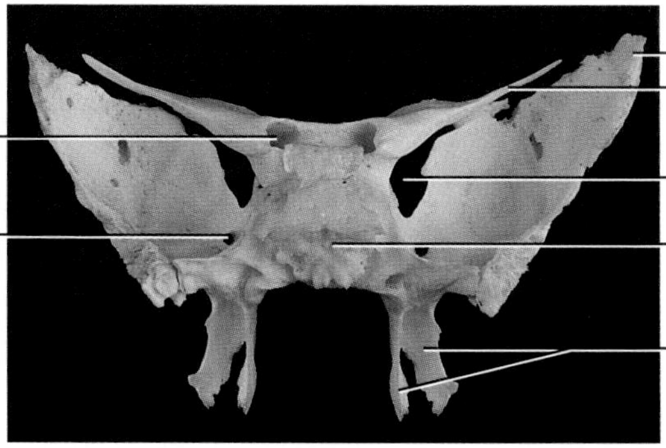

SUPERIOR

Greater wing
Lesser wing

Optic foramen
Superior orbital fissure

Foramen rotundum
Body

Pterygoid processes

INFERIOR

(c) Posterior view

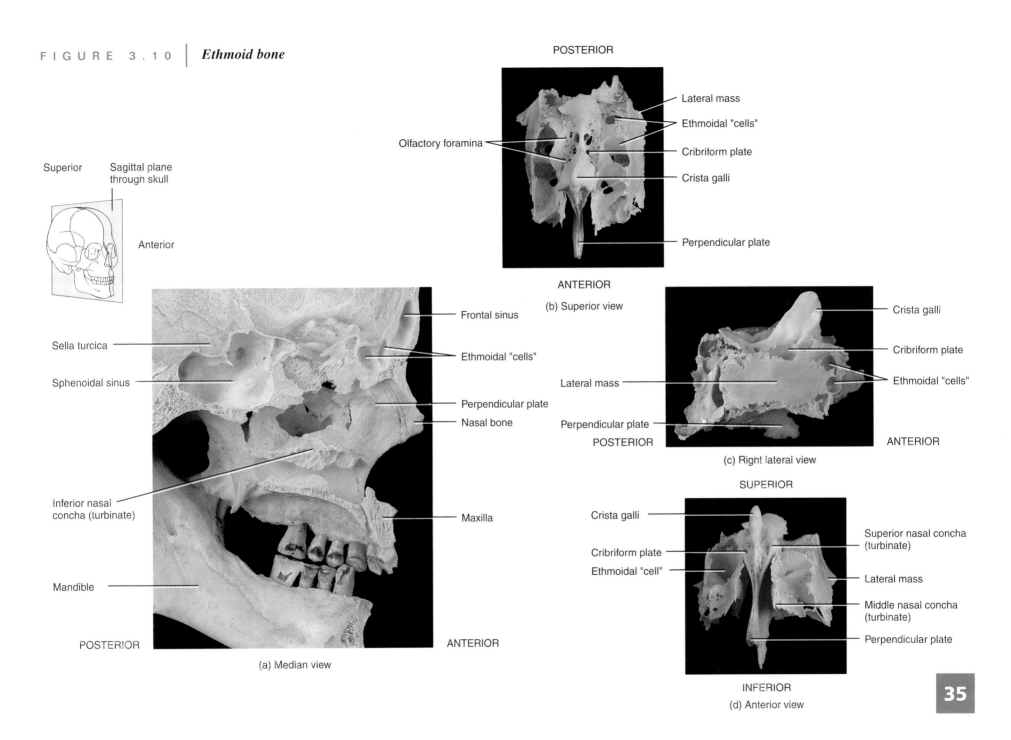

FIGURE 3.10 | *Ethmoid bone*

Superior

Sagittal plane through skull

Anterior

POSTERIOR

Lateral mass

Ethmoidal "cells"

Olfactory foramina

Cribriform plate

Crista galli

Perpendicular plate

ANTERIOR

(b) Superior view

Sella turcica

Frontal sinus

Ethmoidal "cells"

Sphenoidal sinus

Perpendicular plate

Nasal bone

Inferior nasal concha (turbinate)

Maxilla

Mandible

POSTERIOR

ANTERIOR

(a) Median view

Crista galli

Cribriform plate

Lateral mass

Ethmoidal "cells"

Perpendicular plate

POSTERIOR

ANTERIOR

(c) Right lateral view

SUPERIOR

Crista galli

Superior nasal concha (turbinate)

Cribriform plate

Ethmoidal "cell"

Lateral mass

Middle nasal concha (turbinate)

Perpendicular plate

INFERIOR

(d) Anterior view

35

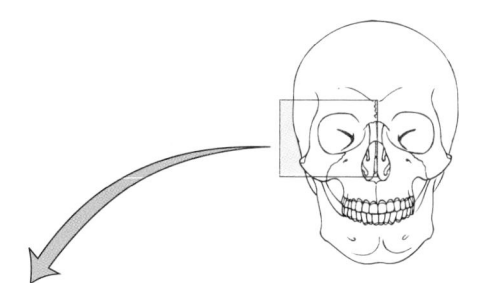

SUPERIOR

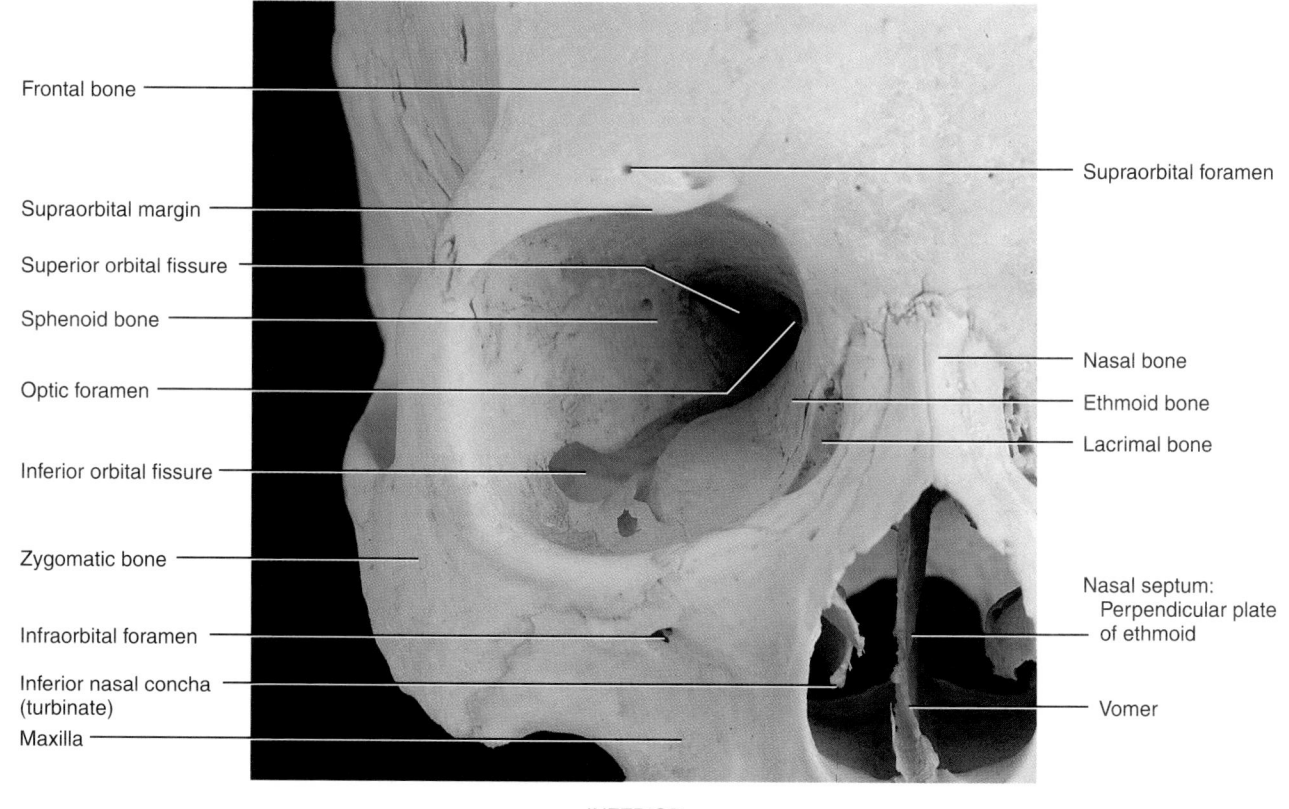

Frontal bone

Supraorbital foramen

Supraorbital margin

Superior orbital fissure

Sphenoid bone

Nasal bone

Optic foramen

Ethmoid bone

Lacrimal bone

Inferior orbital fissure

Zygomatic bone

Nasal septum:
Perpendicular plate
of ethmoid

Infraorbital foramen

Inferior nasal concha
(turbinate)

Maxilla

Vomer

INFERIOR

Anterior view

FIGURE 3.11 | *Right orbit*

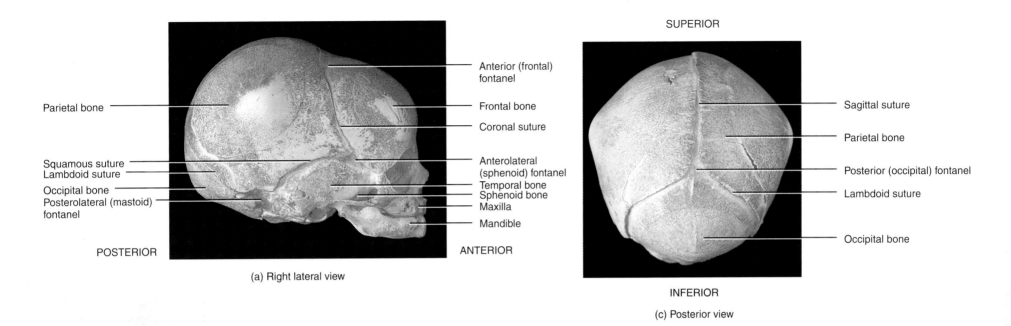

Parietal bone

Anterior (frontal) fontanel

Frontal bone

Coronal suture

Squamous suture
Lambdoid suture

Anterolateral (sphenoid) fontanel

Occipital bone
Posterolateral (mastoid) fontanel

Temporal bone
Sphenoid bone
Maxilla

Mandible

POSTERIOR

ANTERIOR

(a) Right lateral view

SUPERIOR

Sagittal suture

Parietal bone

Posterior (occipital) fontanel

Lambdoid suture

Occipital bone

INFERIOR

(c) Posterior view

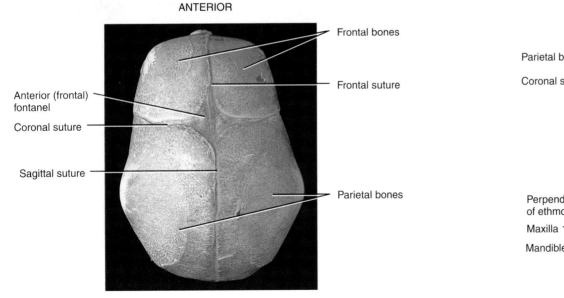

ANTERIOR

Frontal bones

Frontal suture

Anterior (frontal) fontanel

Coronal suture

Sagittal suture

Parietal bones

POSTERIOR

(b) Superior view

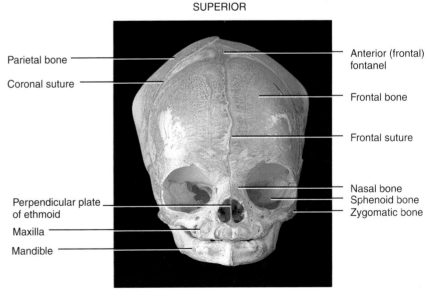

SUPERIOR

Parietal bone

Coronal suture

Anterior (frontal) fontanel

Frontal bone

Frontal suture

Perpendicular plate of ethmoid

Maxilla

Mandible

Nasal bone
Sphenoid bone
Zygomatic bone

INFERIOR

(d) Anterior view

FIGURE 3.12 | *Fontanels of a fetal skull*

37

FIGURE 3.14 | *Auditory ossicles*

INFERIOR

MEDIAL

LATERAL

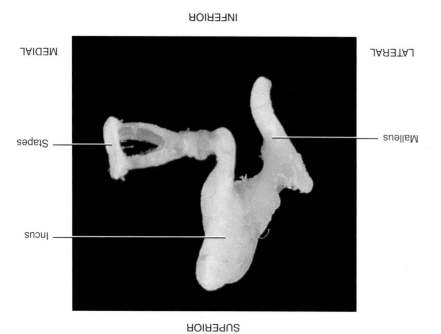

Stapes

Malleus

Incus

SUPERIOR

FIGURE 3.13 | *Hyoid bone*

Superior view

ANTERIOR

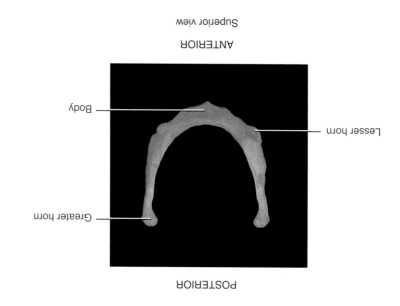

Body

Lesser horn

Greater horn

POSTERIOR

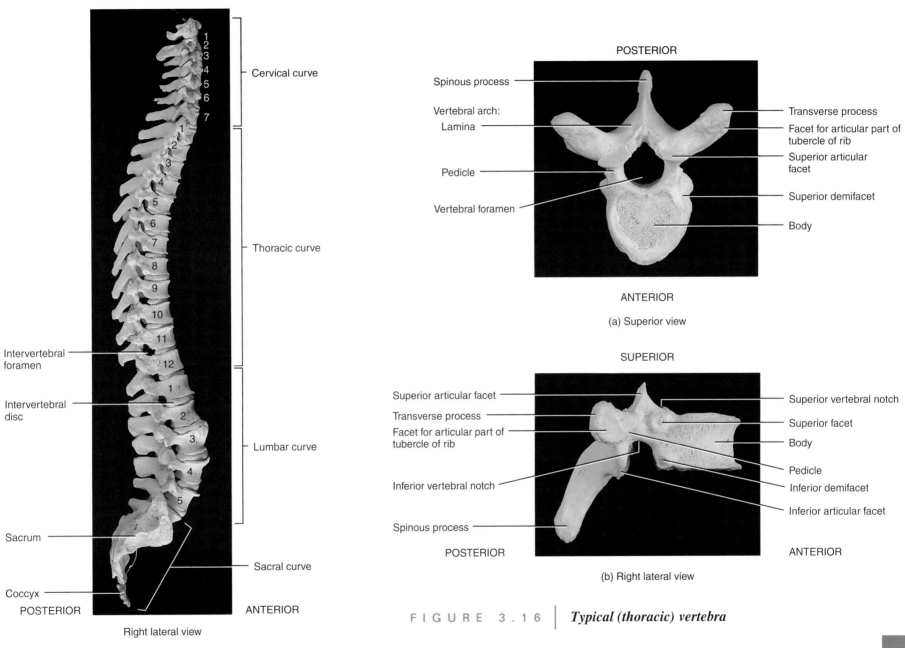

Cervical curve

Thoracic curve

Lumbar curve

Intervertebral foramen

Intervertebral disc

Sacrum

Sacral curve

Coccyx

POSTERIOR

ANTERIOR

Right lateral view

FIGURE 3.15 | *Vertebral column*

POSTERIOR

Spinous process

Vertebral arch:
Lamina

Pedicle

Vertebral foramen

Transverse process

Facet for articular part of tubercle of rib

Superior articular facet

Superior demifacet

Body

ANTERIOR

(a) Superior view

SUPERIOR

Superior articular facet

Transverse process

Facet for articular part of tubercle of rib

Inferior vertebral notch

Spinous process

POSTERIOR

Superior vertebral notch

Superior facet

Body

Pedicle

Inferior demifacet

Inferior articular facet

ANTERIOR

(b) Right lateral view

FIGURE 3.16 | *Typical (thoracic) vertebra*

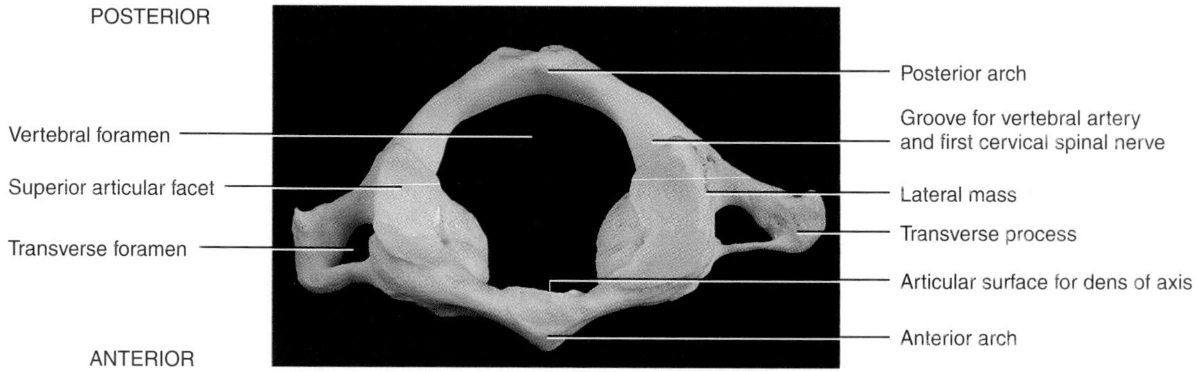

POSTERIOR

Vertebral foramen

Superior articular facet

Transverse foramen

ANTERIOR

Posterior arch

Groove for vertebral artery and first cervical spinal nerve

Lateral mass

Transverse process

Articular surface for dens of axis

Anterior arch

(a) Superior view of the atlas (C1)

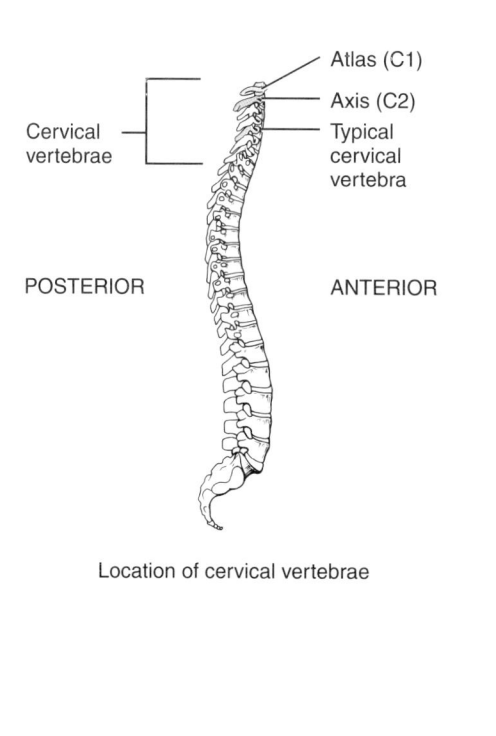

Atlas (C1)

Axis (C2)

Typical cervical vertebra

Cervical vertebrae

POSTERIOR          ANTERIOR

Location of cervical vertebrae

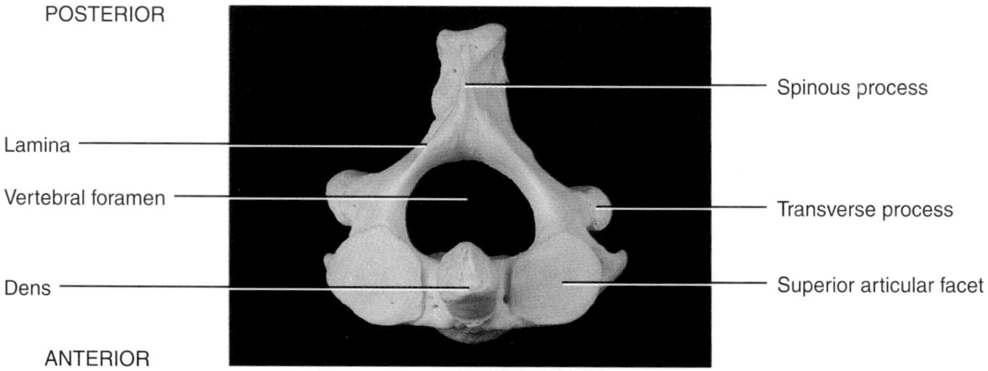

POSTERIOR

Lamina

Vertebral foramen

Dens

ANTERIOR

Spinous process

Transverse process

Superior articular facet

(b) Superior view of the axis (C2)

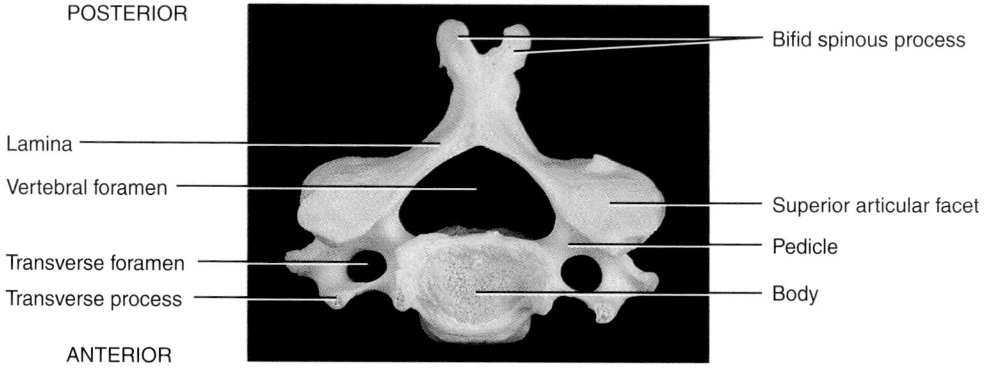

POSTERIOR

Lamina

Vertebral foramen

Transverse foramen

Transverse process

ANTERIOR

Bifid spinous process

Superior articular facet

Pedicle

Body

(c) Superior view of a typical cervical vertebra

FIGURE 3.17 | *Cervical vertebrae*

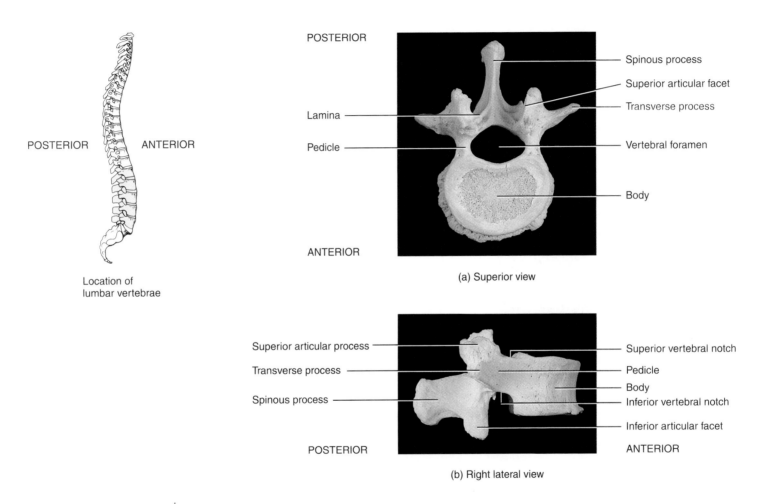

POSTERIOR

POSTERIOR          ANTERIOR

Location of
lumbar vertebrae

POSTERIOR

Lamina

Pedicle

ANTERIOR

Spinous process

Superior articular facet

Transverse process

Vertebral foramen

Body

(a) Superior view

Superior articular process

Transverse process

Spinous process

Superior vertebral notch

Pedicle

Body

Inferior vertebral notch

Inferior articular facet

POSTERIOR                ANTERIOR

(b) Right lateral view

FIGURE 3.18 | *Lumbar vertebrae*

SUPERIOR

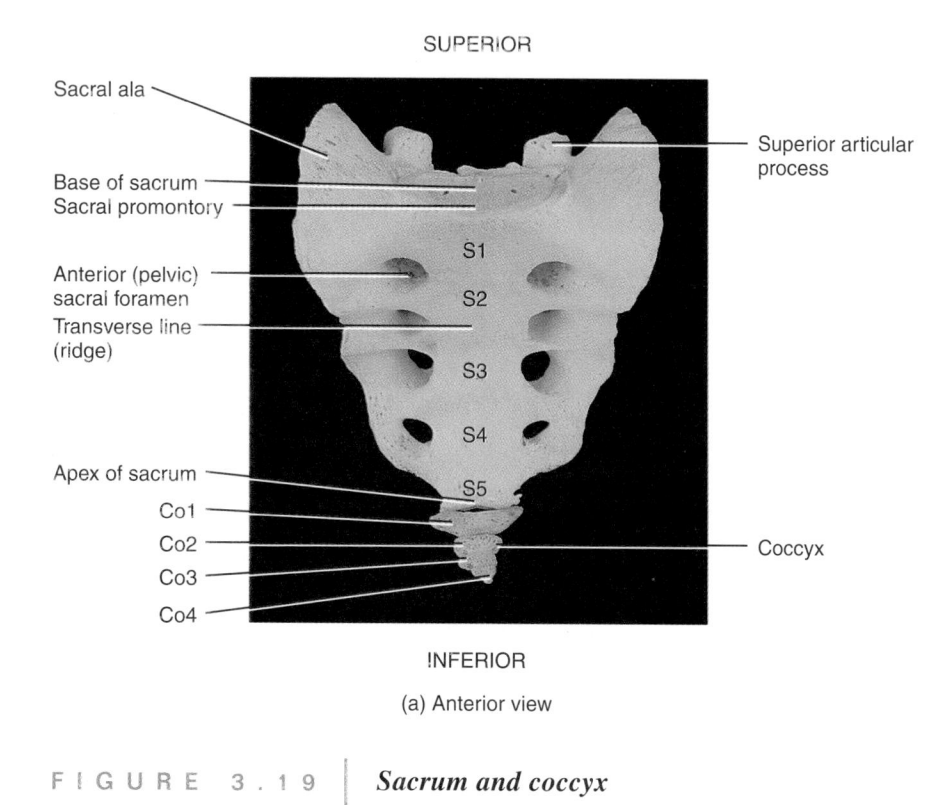

Sacral ala

Base of sacrum

Sacral promontory

Anterior (pelvic)
sacral foramen

Transverse line
(ridge)

Apex of sacrum

Co1

Co2

Co3

Co4

Superior articular
process

S1

S2

S3

S4

S5

Coccyx

INFERIOR

(a) Anterior view

SUPERIOR

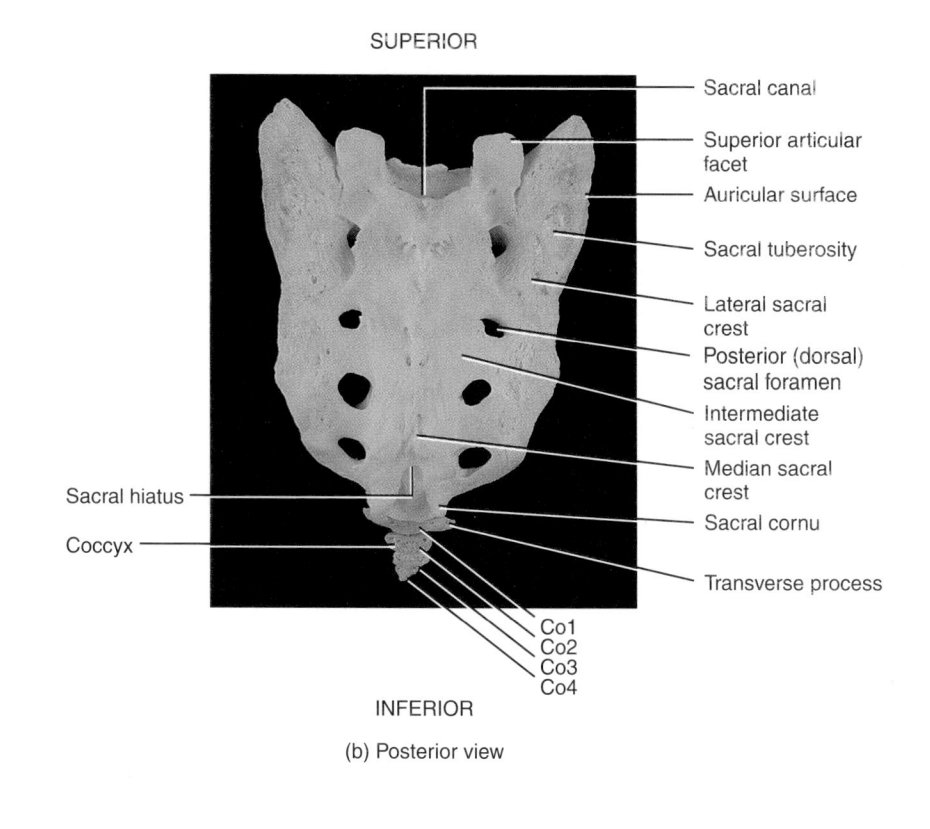

Sacral canal

Superior articular
facet

Auricular surface

Sacral tuberosity

Lateral sacral
crest

Posterior (dorsal)
sacral foramen

Intermediate
sacral crest

Median sacral
crest

Sacral cornu

Transverse process

Sacral hiatus

Coccyx

Co1
Co2
Co3
Co4

INFERIOR

(b) Posterior view

FIGURE 3.19 | *Sacrum and coccyx*

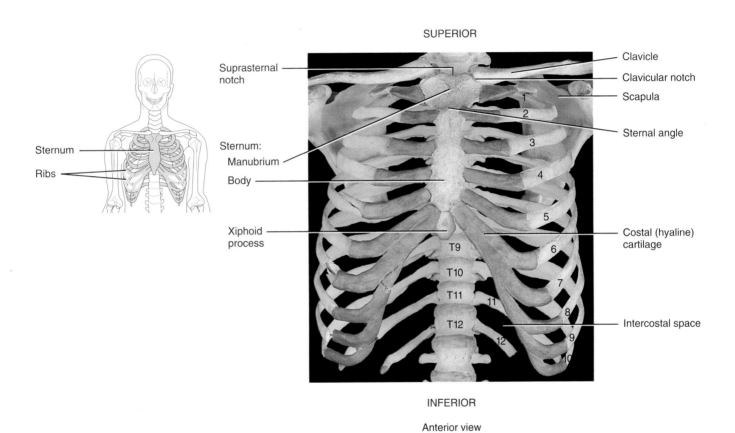

SUPERIOR

Clavicle

Clavicular notch

Scapula

Sternal angle

Suprasternal notch

Sternum:
  Manubrium

Body

Xiphoid process

Costal (hyaline) cartilage

Intercostal space

Sternum

Ribs

1
2
3
4
5
6
7
8
9
10
11
12

T9
T10
T11
T12

INFERIOR

Anterior view

FIGURE  3.20 | *Anterior view of skeleton of thorax*

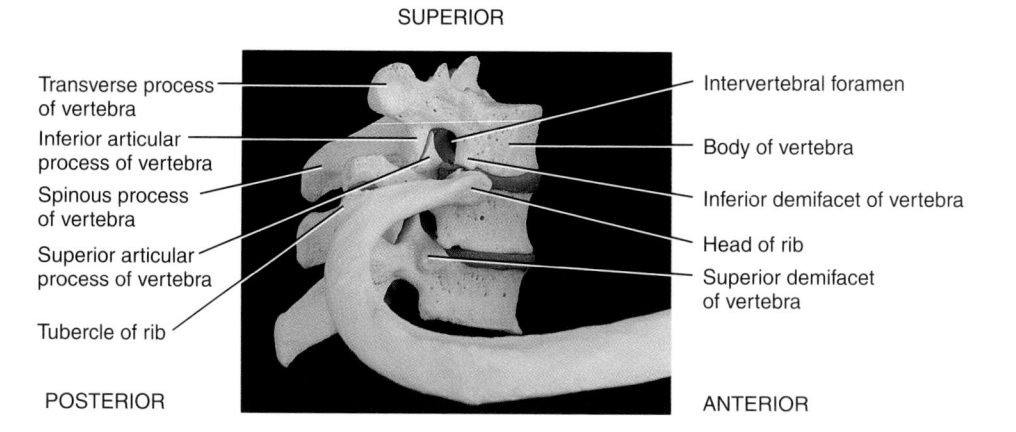

SUPERIOR

Transverse process of vertebra

Inferior articular process of vertebra

Spinous process of vertebra

Superior articular process of vertebra

Tubercle of rib

Intervertebral foramen

Body of vertebra

Inferior demifacet of vertebra

Head of rib

Superior demifacet of vertebra

POSTERIOR

ANTERIOR

(a) Right lateral view

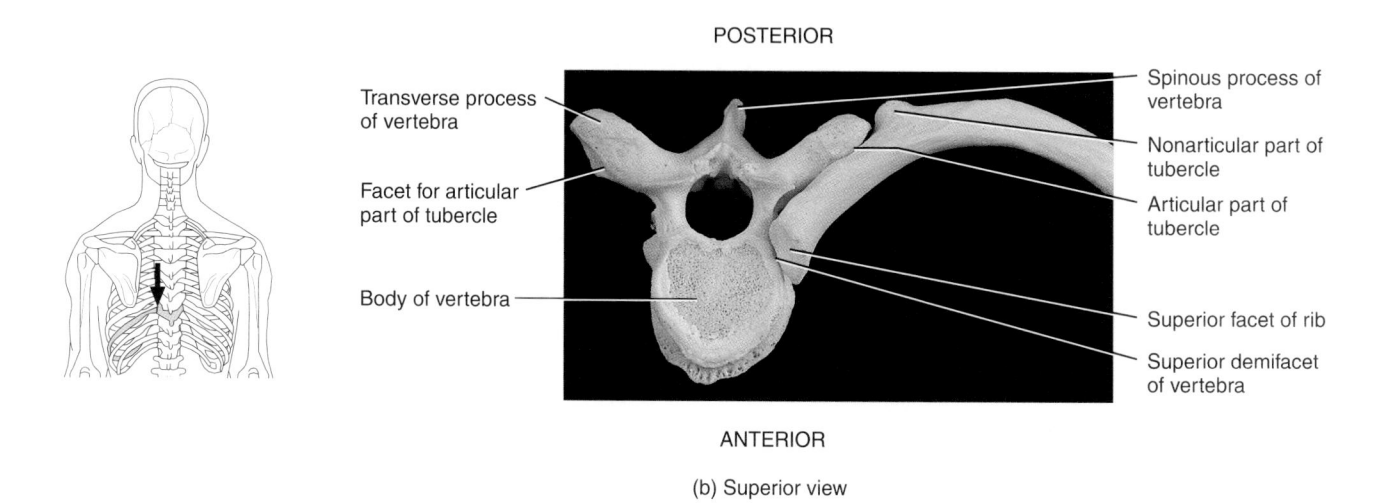

POSTERIOR

Transverse process of vertebra

Facet for articular part of tubercle

Body of vertebra

Spinous process of vertebra

Nonarticular part of tubercle

Articular part of tubercle

Superior facet of rib

Superior demifacet of vertebra

ANTERIOR

(b) Superior view

FIGURE 3.21 | *Articulation of a thoracic vertebra with a rib*

FIGURE 3.22 | *Right scapula*

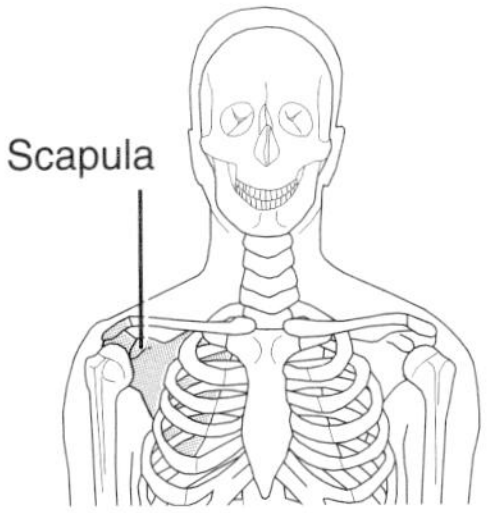

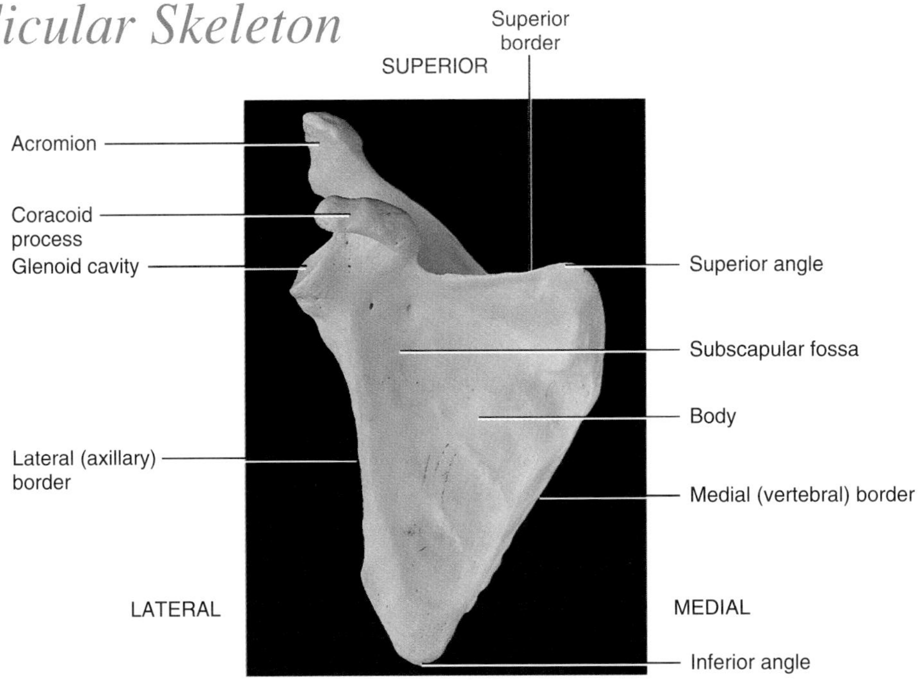

# *Appendicular Skeleton*

Scapula

SUPERIOR

Superior border

Acromion

Coracoid process

Glenoid cavity

Superior angle

Subscapular fossa

Body

Lateral (axillary) border

Medial (vertebral) border

LATERAL

MEDIAL

Inferior angle

(a) Anterior view

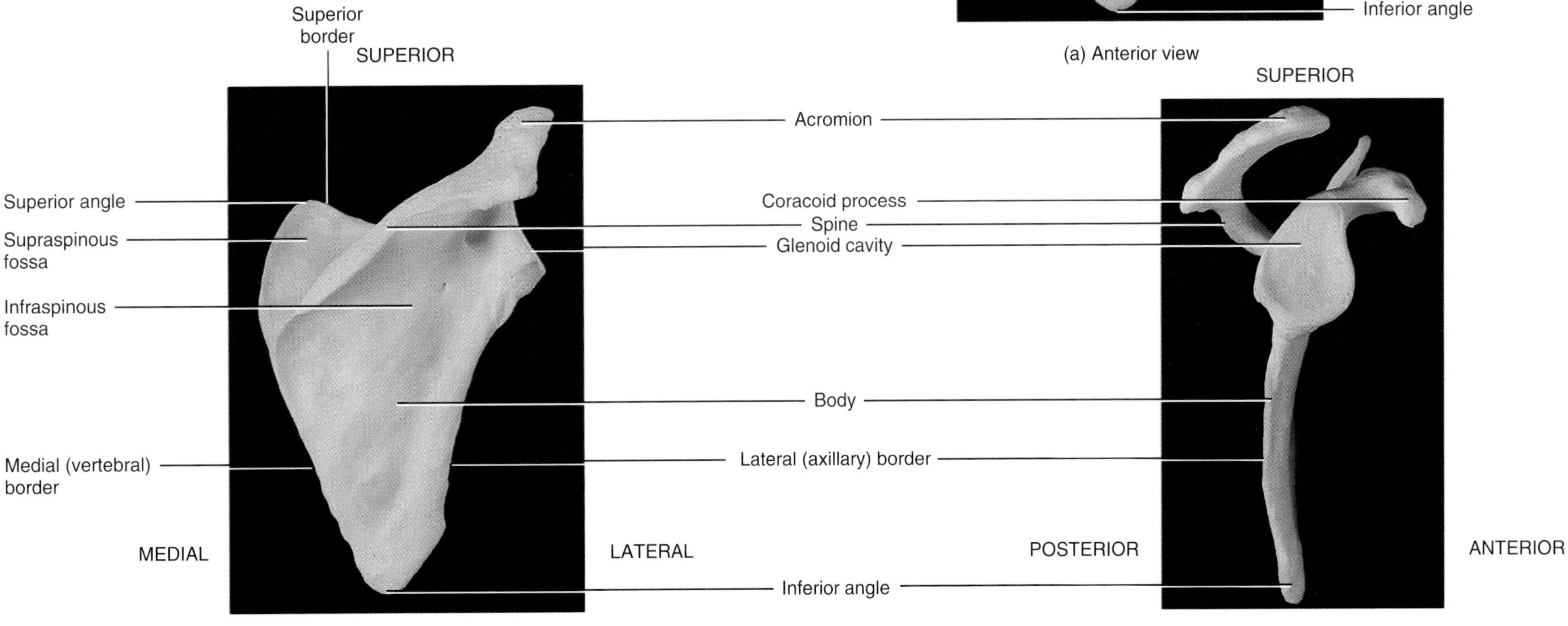

Superior border
SUPERIOR

Superior angle

Supraspinous fossa

Infraspinous fossa

Acromion

Coracoid process
Spine
Glenoid cavity

Body

Medial (vertebral) border

Lateral (axillary) border

MEDIAL

LATERAL

Inferior angle

(b) Posterior view

SUPERIOR

POSTERIOR

ANTERIOR

(c) Lateral border view

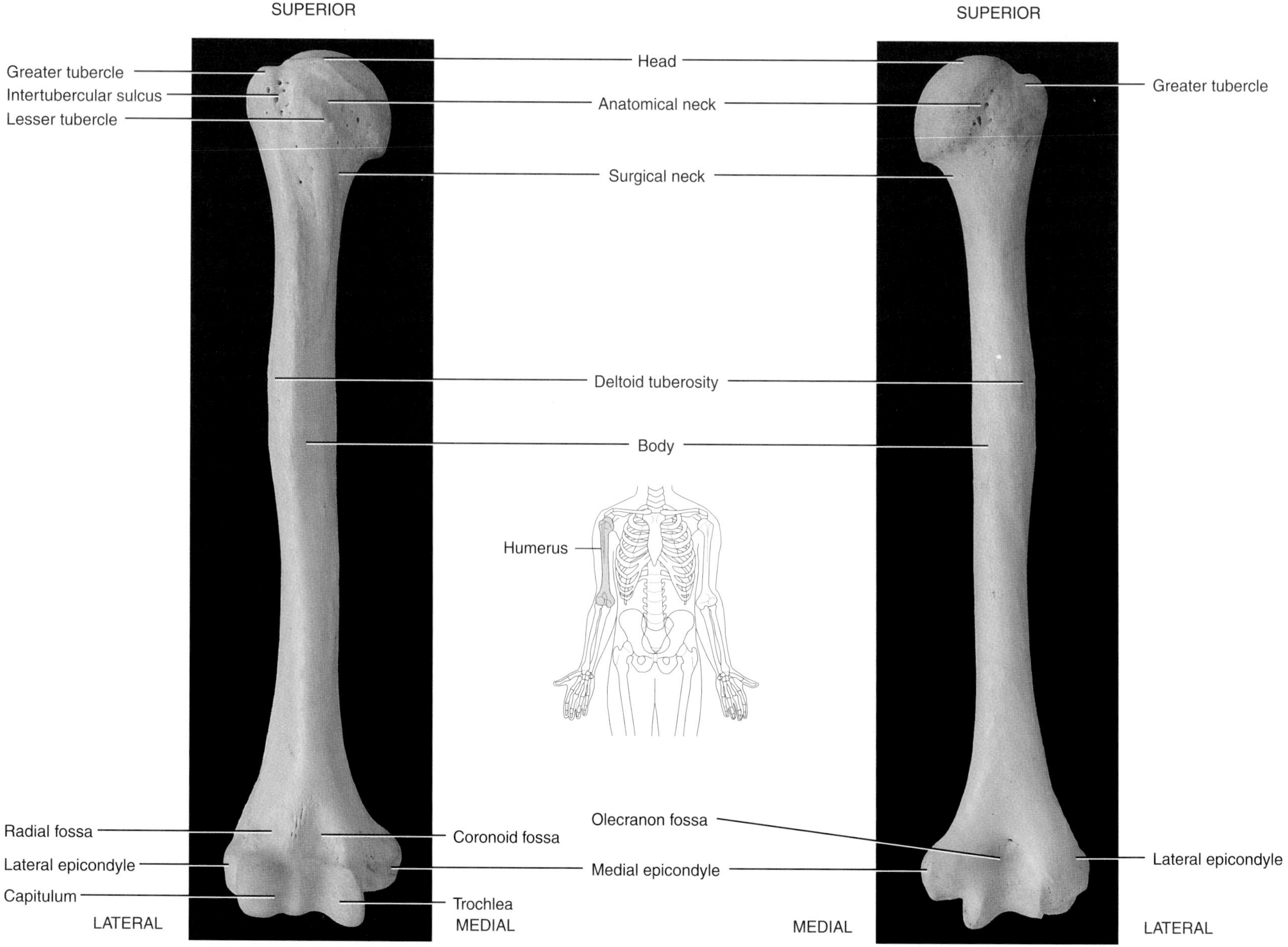

SUPERIOR

Greater tubercle
Intertubercular sulcus
Lesser tubercle

SUPERIOR

Head
Anatomical neck

Greater tubercle

Surgical neck

Deltoid tuberosity

Body

Humerus

Radial fossa
Lateral epicondyle
Capitulum

Coronoid fossa

Olecranon fossa

Medial epicondyle

Lateral epicondyle

Trochlea
LATERAL    MEDIAL

MEDIAL    LATERAL

(a) Anterior view

(b) Posterior view

FIGURE 3.23    *Right humerus*

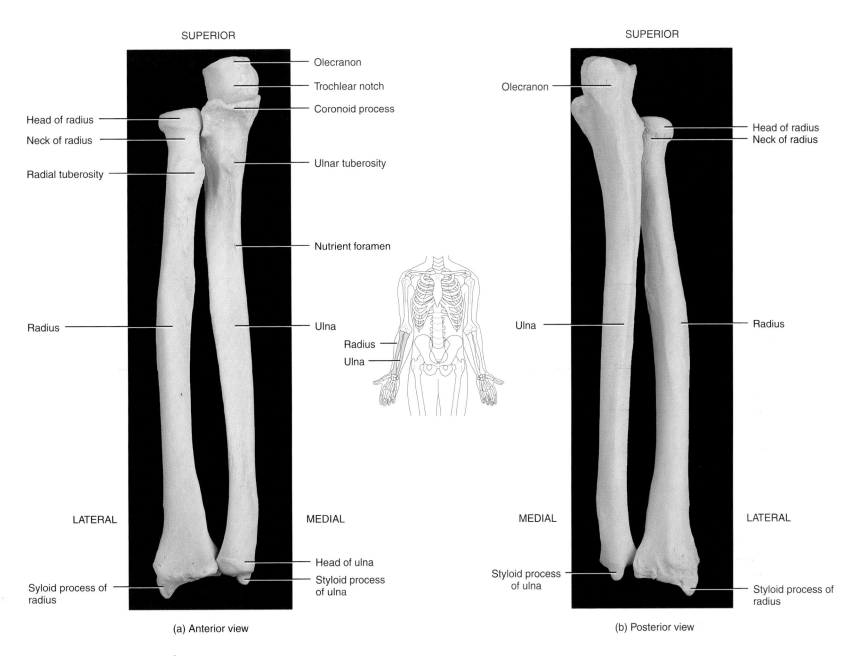

SUPERIOR

Olecranon

Trochlear notch

Coronoid process

Head of radius

Neck of radius

Radial tuberosity

Ulnar tuberosity

Nutrient foramen

Radius

Ulna

Radius

Ulna

LATERAL

MEDIAL

Syloid process of
radius

Head of ulna

Styloid process
of ulna

(a) Anterior view

SUPERIOR

Olecranon

Head of radius
Neck of radius

Ulna

Radius

MEDIAL

LATERAL

Styloid process
of ulna

Styloid process of
radius

(b) Posterior view

FIGURE  3 . 2 4   |   *Right ulna and radius*

47

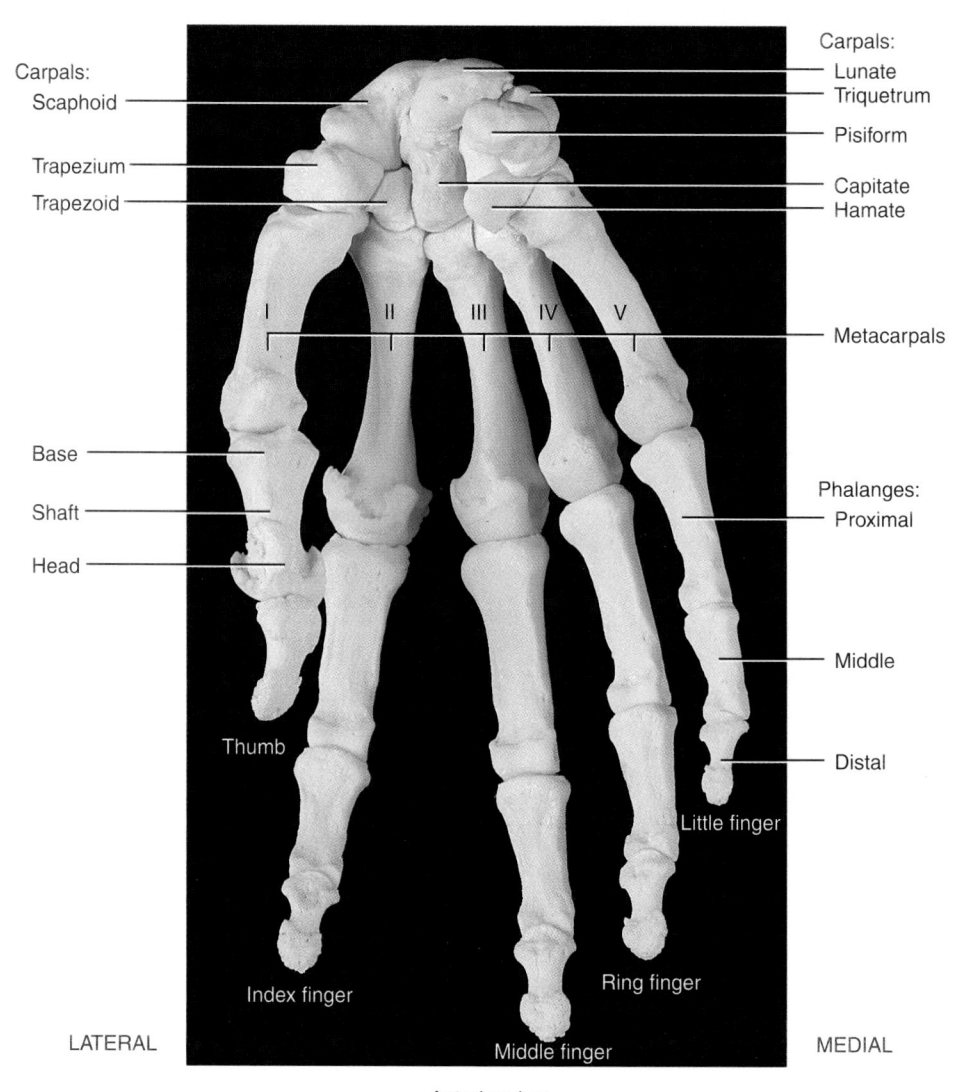

SUPERIOR

Carpals

Metacarpals

Phalanges

Carpals:
Scaphoid

Trapezium

Trapezoid

Base

Shaft

Head

Thumb

Index finger

Carpals:
Lunate
Triquetrum
Pisiform

Capitate
Hamate

I    II    III    IV    V

Metacarpals

Phalanges:
Proximal

Middle

Distal

Little finger

Ring finger

Middle finger

LATERAL

MEDIAL

Anterior view

FIGURE  3.25  |  *Right hand and wrist*

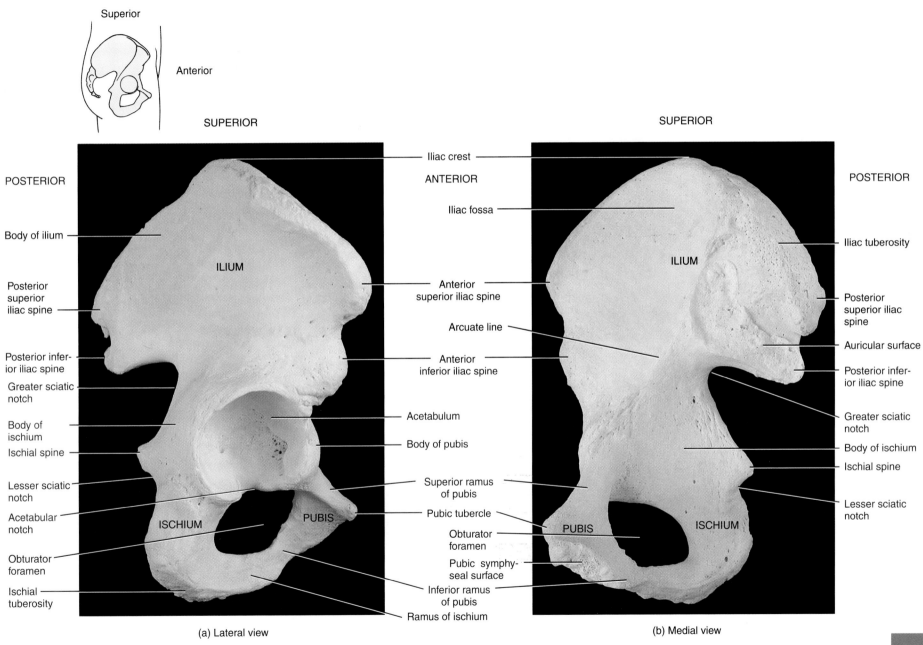

Superior

Anterior

**SUPERIOR**

**SUPERIOR**

POSTERIOR

Iliac crest

POSTERIOR

ANTERIOR

Iliac fossa

Body of ilium

ILIUM

Iliac tuberosity

ILIUM

Posterior
superior
iliac spine

Anterior
superior iliac spine

Posterior
superior iliac
spine

Arcuate line

Auricular surface

Posterior infer-
ior iliac spine

Anterior
inferior iliac spine

Posterior infer-
ior iliac spine

Greater sciatic
notch

Acetabulum

Greater sciatic
notch

Body of
ischium

Body of pubis

Body of ischium

Ischial spine

Ischial spine

Superior ramus
of pubis

Lesser sciatic
notch

Lesser sciatic
notch

Acetabular
notch

ISCHIUM

PUBIS

Pubic tubercle

PUBIS

ISCHIUM

Obturator
foramen

Obturator
foramen

Pubic symphy-
seal surface

Ischial
tuberosity

Inferior ramus
of pubis

Ramus of ischium

(a) Lateral view

(b) Medial view

F I G U R E   3 . 2 6   |   *Right hip bone*

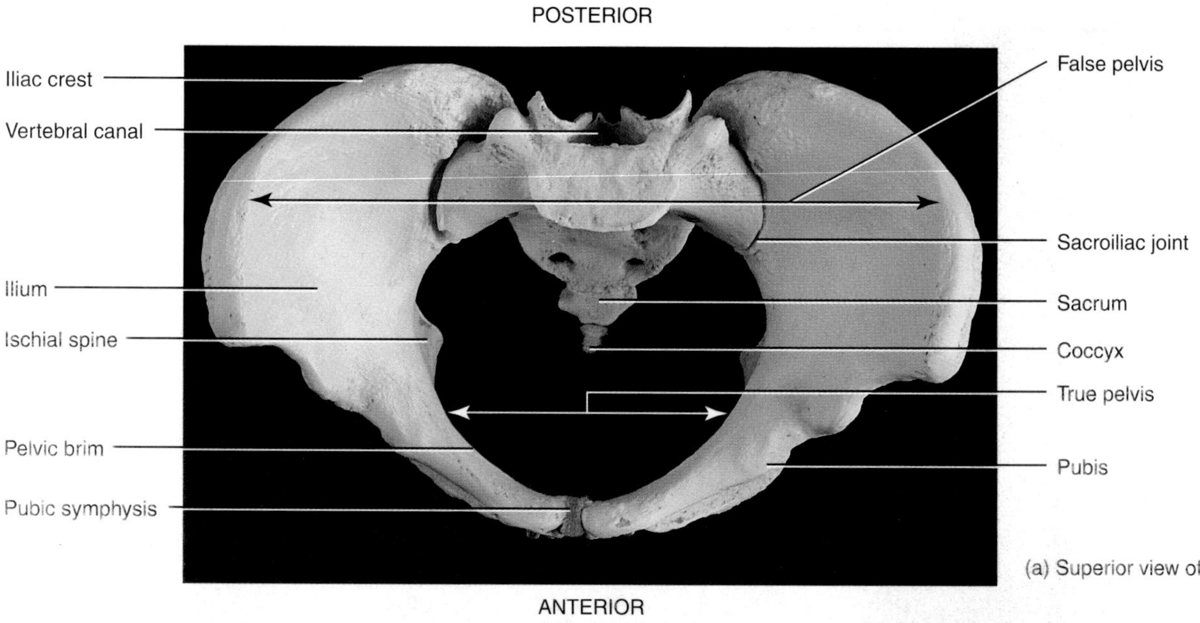

POSTERIOR

Iliac crest

Vertebral canal

Ilium

Ischial spine

Pelvic brim

Pubic symphysis

False pelvis

Sacroiliac joint

Sacrum

Coccyx

True pelvis

Pubis

(a) Superior view of female pelvis

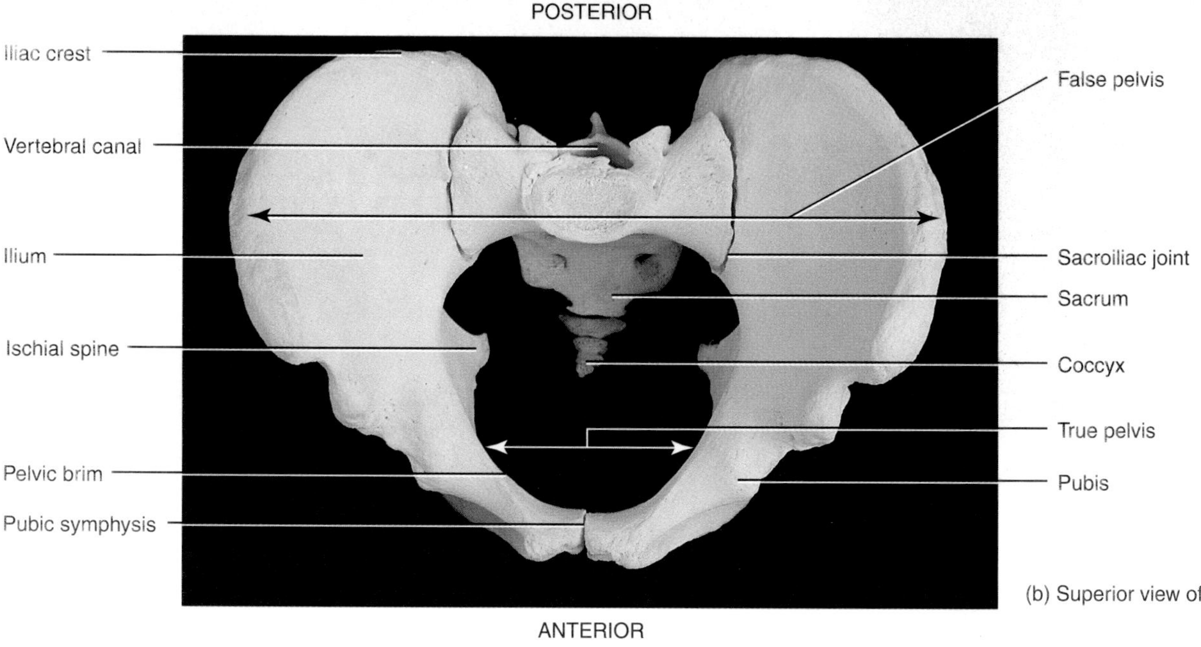

ANTERIOR

POSTERIOR

Iliac crest

Vertebral canal

Ilium

Ischial spine

Pelvic brim

Pubic symphysis

False pelvis

Sacroiliac joint

Sacrum

Coccyx

True pelvis

Pubis

(b) Superior view of male pelvis

ANTERIOR

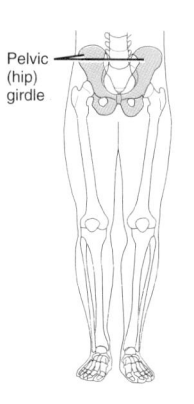

Pelvic (hip) girdle

FIGURE 3.27 | *Pelvis*

FIGURE 3.28 **Right femur**

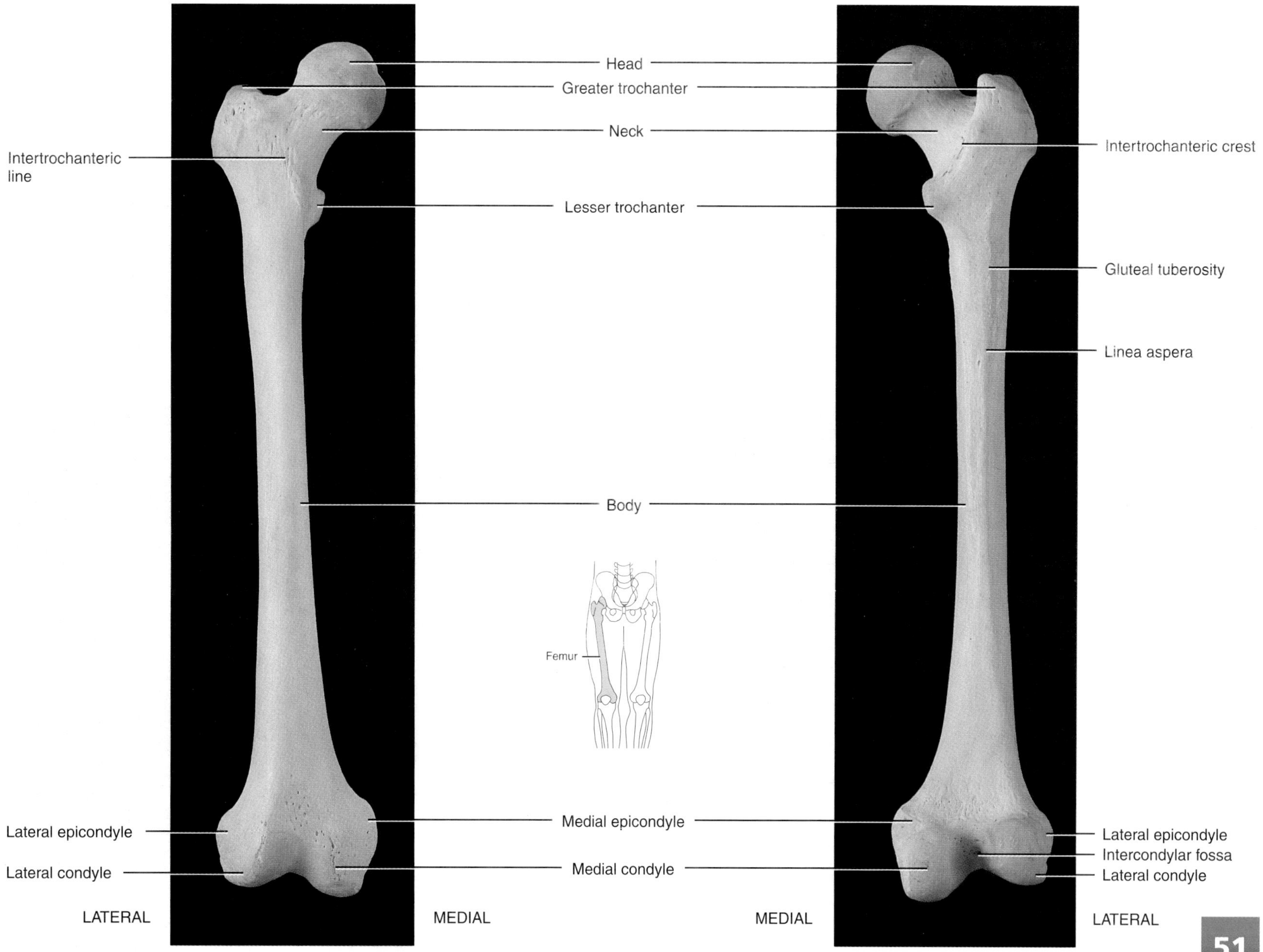

SUPERIOR

Head

Greater trochanter

Neck

Intertrochanteric line

Lesser trochanter

Intertrochanteric crest

Gluteal tuberosity

Linea aspera

Body

Femur

Lateral epicondyle

Lateral condyle

Medial epicondyle

Medial condyle

SUPERIOR

Lateral epicondyle

Intercondylar fossa

Lateral condyle

LATERAL

MEDIAL

MEDIAL

LATERAL

(a) Anterior view

(b) Posterior view

51

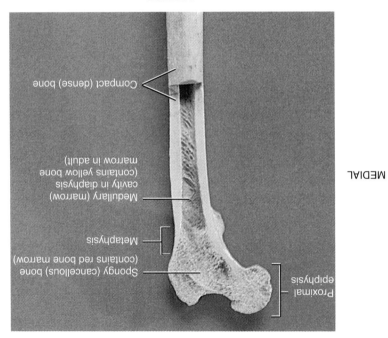

FIGURE 3.29 | *Portion of a partially sectioned femur*

INFERIOR

Anterior view

Compact (dense) bone

LATERAL

MEDIAL

Medullary (marrow) cavity in diaphysis (contains yellow bone marrow in adult)

Metaphysis

Spongy (cancellous) bone (contains red bone marrow)

Proximal epiphysis

SUPERIOR

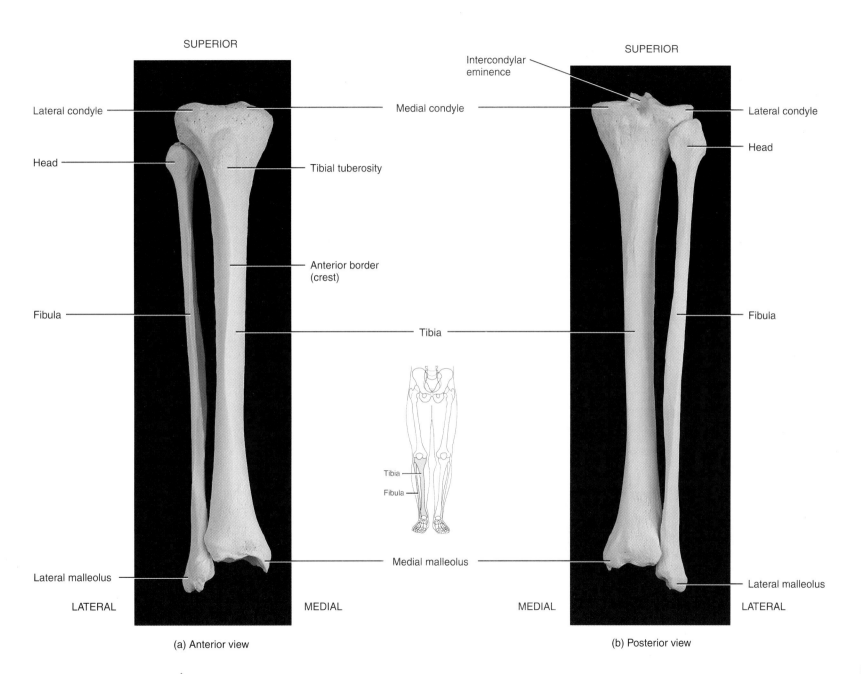

SUPERIOR

Lateral condyle

Head

Fibula

Anterior border (crest)

Tibia

Medial condyle

Tibial tuberosity

Medial malleolus

Lateral malleolus

LATERAL

MEDIAL

(a) Anterior view

SUPERIOR

Intercondylar eminence

Lateral condyle

Head

Fibula

Tibia

Lateral malleolus

MEDIAL

LATERAL

(b) Posterior view

Tibia

Fibula

FIGURE 3.30 | *Right tibia and fibula*

53

54

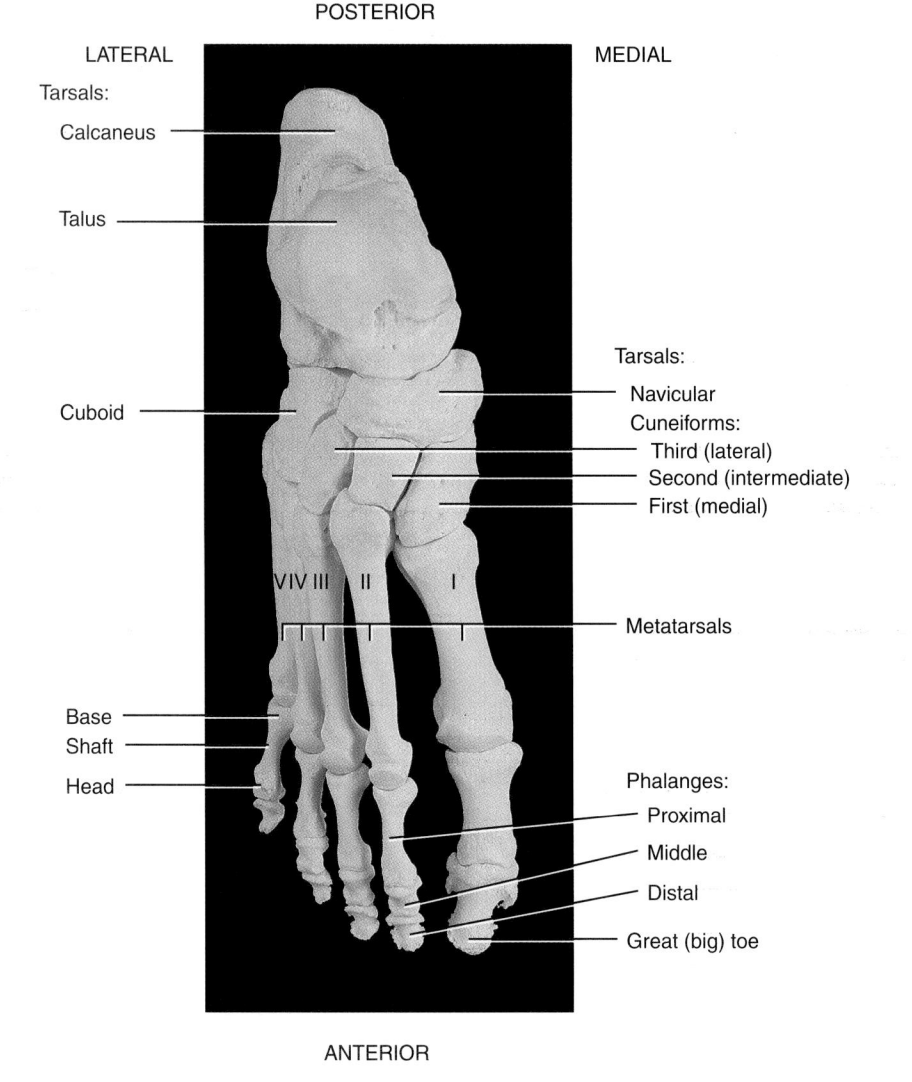
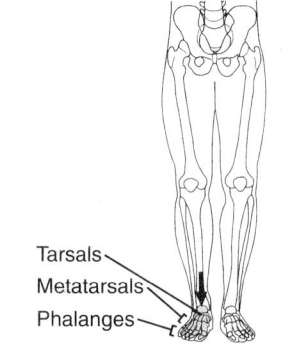

POSTERIOR

LATERAL    MEDIAL

Tarsals:

Calcaneus

Talus

Cuboid

Tarsals:

Navicular

Cuneiforms:

Third (lateral)

Second (intermediate)

First (medial)

V IV III    II    I

Metatarsals

Base

Shaft

Head

Phalanges:

Proximal

Middle

Distal

Great (big) toe

ANTERIOR

Superior view

Tarsals
Metatarsals
Phalanges

FIGURE 3.31 | *Right foot and ankle*

**T A B L E   4 . 1** | ***Summary of Structural Categories and Functional Characteristics of Joints***

| Structural Category | Description | Functional Characteristics | Example |
|---|---|---|---|
| **FIBROUS JOINTS:** | Articulating bones held together by fibrous connective tissue; no synovial cavity. | | |
| Suture | Articulating bones united by a thin layer of dense fibrous connective tissue; found between bones of the skull. With age, some sutures are replaced by synostosis, in which the bones fuse across the former suture. | Synarthrosis (immovable). | Frontal suture. |
| Syndesmosis | Articulating bones are united by a dense fibrous connective tissue, either a ligament or an interosseous membrane. | Amphiarthrosis (slightly movable). | Distal tibiofibular joint. |
| Gomphosis | Articulating bones united by a periodontal ligament; cone-shaped peg fits into a socket. | Synarthrosis. | Roots of teeth in alveoli (sockets) of maxillae and mandible. |
| **CARTILAGINOUS JOINTS:** | Articulating bones united by cartilage; no synovial cavity. | | |
| Synchondrosis | Connecting material is hyaline cartilage; becomes a synostosis when bone elongation ceases. | Synarthrosis. | Epiphyseal plate at joint between the diaphysis and epiphysis of a long bone. |
| Symphysis | Connecting material is a broad, flat disc of fibrocartilage. | Amphiarthrosis. | Intervertebral joints, pubic symphysis, and junction of manubrium with body of sternum. |
| **SYNOVIAL JOINTS:** | Characterized by a synovial cavity, articular cartilage, and an articular capsule; may contain accessory ligaments, articular discs, and bursae. | | |
| Planar | Articulating surfaces are flat or just slightly curved. | Nonaxial diarthrosis (freely movable); gliding motion. | Intercarpal, intertarsal, sternocostal (between sternum and the 2nd–7th pairs of ribs), and vertebrocostal joints. |
| Hinge | Convex surface fits into a concave surface. | Monaxial diarthrosis; angular motion. | Elbow, ankle, and interphalangeal joints. |
| Pivot | Rounded or pointed surface fits into a ring formed partly by bone and partly by a ligament. | Monaxial diarthrosis; rotation. | Atlanto-axial and radioulnar joints. |
| Condyloid | Oval-shaped projection fits into an oval-shaped depression. | Biaxial diarthrosis; angular motion. | Radiocarpal and metacarpophalangeal joints. |
| Saddle | Articular surface of one bone is saddle-shaped, and the articular surface of the other bone "sits" in the saddle. | Biaxial diarthrosis; angular motion. | Carpometacarpal joint between trapezium and thumb. |
| Ball-and-socket | Ball-like surface fits into a cuplike depression. | Multiaxial diarthrosis; angular motion rotation. | Shoulder and hip joints. |

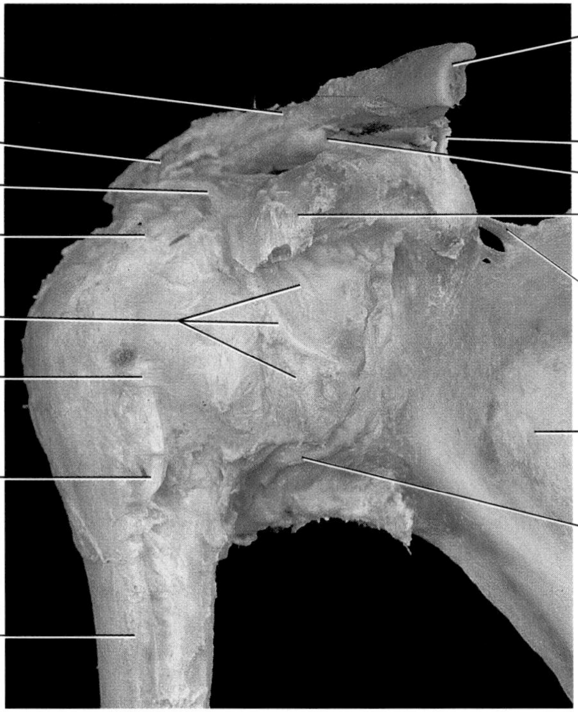

SUPERIOR

Clavicle (cut)

Acromioclavicular
ligament

Acromion of
scapula

Coracoacromial
ligament

Coracohumeral
ligament

Glenohumeral
ligament

Transverse humeral
ligament

Tendon sheath of
biceps brachii
(long head)

Humerus

Coracoclavicular
ligament:
 Conoid ligament

Trapezoid ligament

Coracoid process
of scapula

Superior transverse
scapular ligament

Scapula

Articular capsule

LATERAL

MEDIAL

INFERIOR

Anterior view

FIGURE 4.1 | *Right shoulder (glenohumeral) joint*

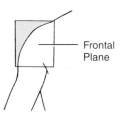

Frontal
Plane

SUPERIOR

LATERAL

MEDIAL

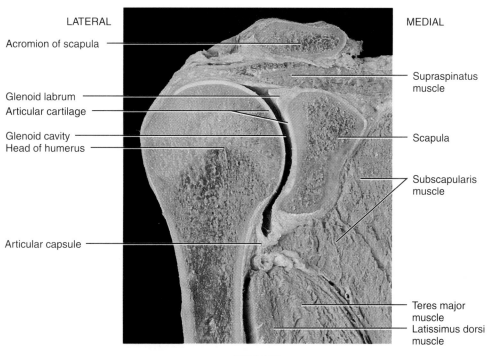

Acromion of scapula

Supraspinatus
muscle

Glenoid labrum

Articular cartilage

Glenoid cavity

Head of humerus

Scapula

Subscapularis
muscle

Articular capsule

Teres major
muscle

Latissimus dorsi
muscle

INFERIOR

Frontal section

FIGURE 4.2 | *Right shoulder (glenohumeral)*
*joint*

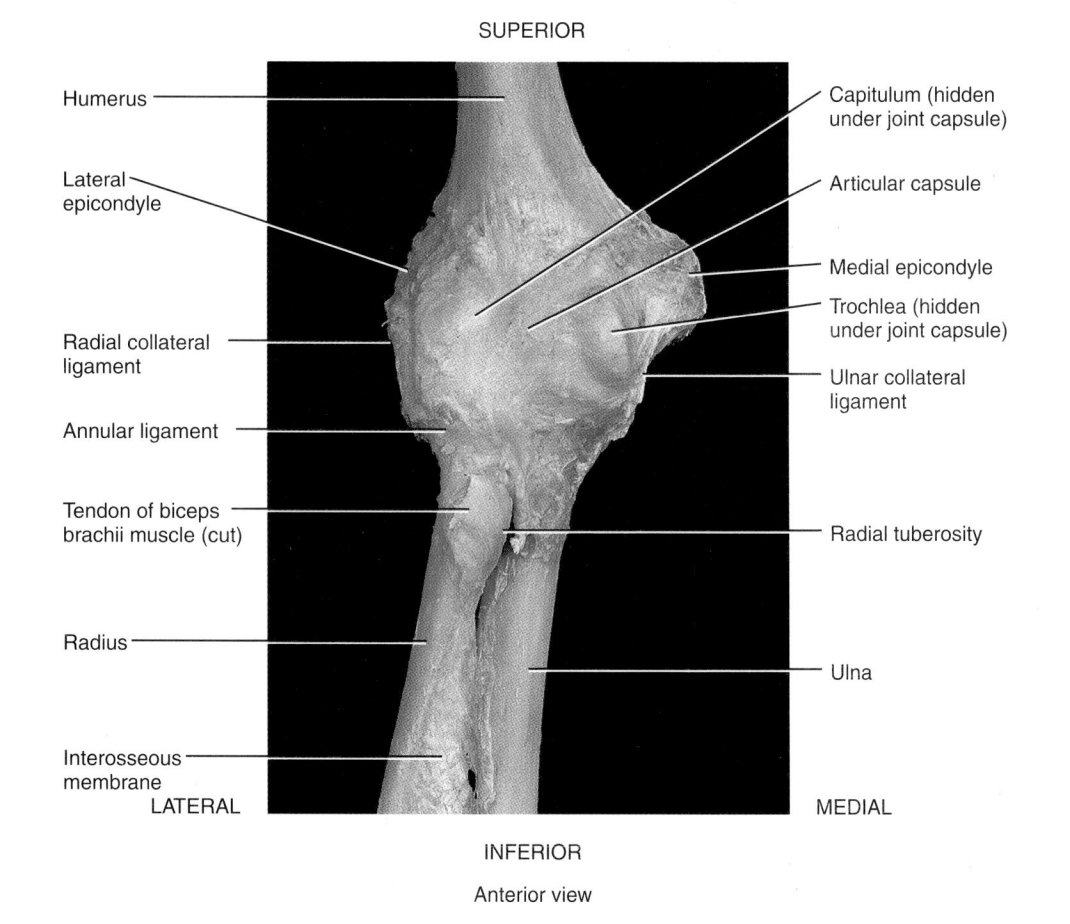

SUPERIOR

Humerus

Lateral
epicondyle

Radial collateral
ligament

Annular ligament

Tendon of biceps
brachii muscle (cut)

Radius

Interosseous
membrane

Capitulum (hidden
under joint capsule)

Articular capsule

Medial epicondyle

Trochlea (hidden
under joint capsule)

Ulnar collateral
ligament

Radial tuberosity

Ulna

LATERAL

MEDIAL

INFERIOR

Anterior view

FIGURE 4.3 | *Right elbow joint*

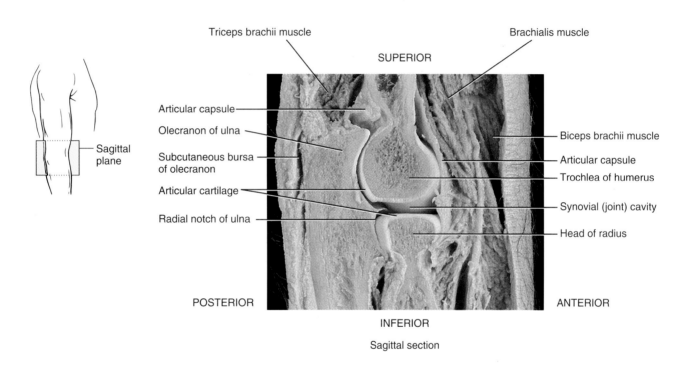

Triceps brachii muscle

Brachialis muscle

SUPERIOR

Articular capsule

Olecranon of ulna

Subcutaneous bursa
of olecranon

Articular cartilage

Radial notch of ulna

Biceps brachii muscle

Articular capsule

Trochlea of humerus

Synovial (joint) cavity

Head of radius

Sagittal
plane

POSTERIOR

ANTERIOR

INFERIOR

Sagittal section

F I G U R E   4 . 4 | *Right elbow joint*

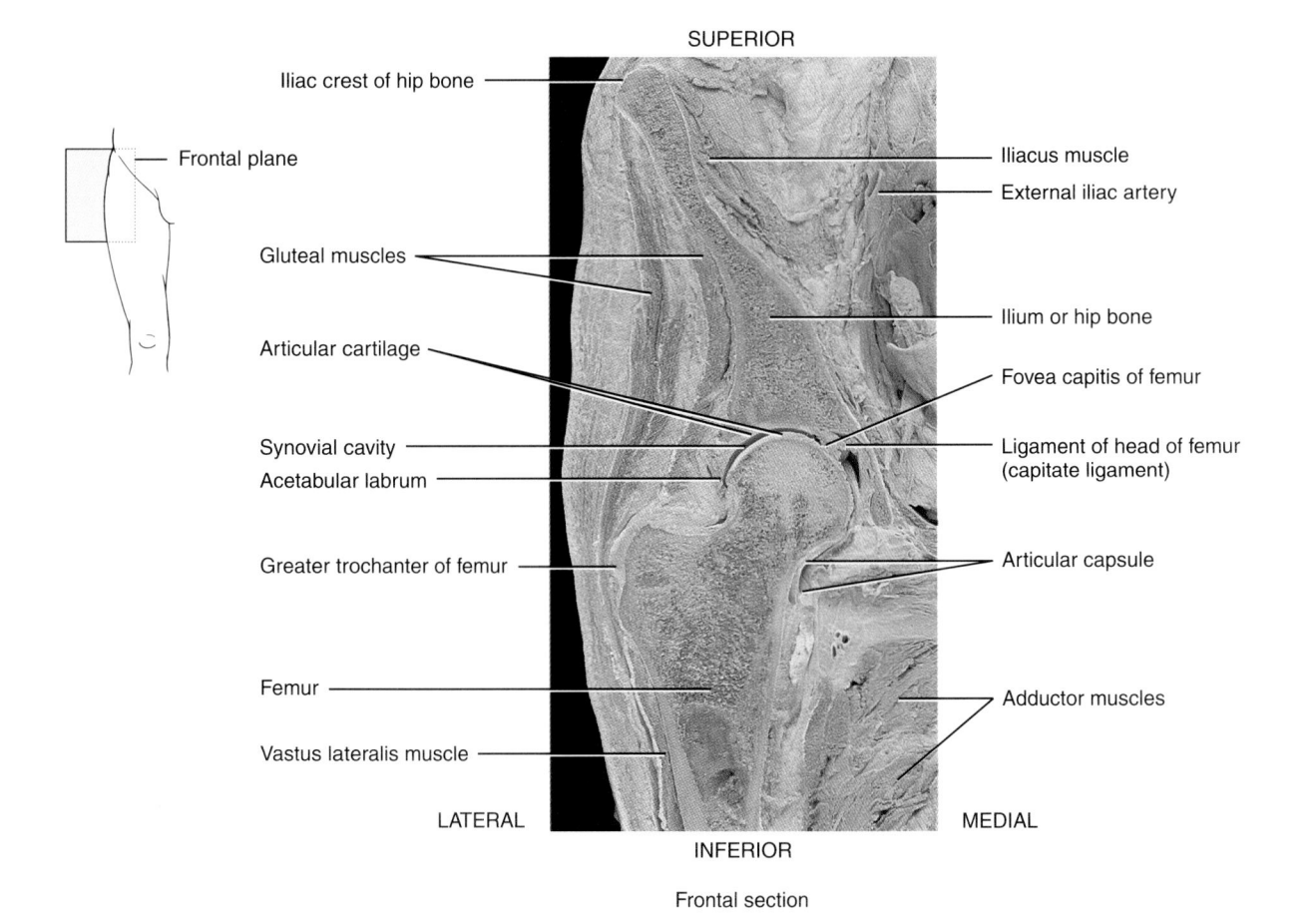

SUPERIOR

Iliac crest of hip bone

Frontal plane

Iliacus muscle

External iliac artery

Gluteal muscles

Ilium or hip bone

Articular cartilage

Fovea capitis of femur

Synovial cavity

Ligament of head of femur
(capitate ligament)

Acetabular labrum

Greater trochanter of femur

Articular capsule

Femur

Adductor muscles

Vastus lateralis muscle

LATERAL

MEDIAL

INFERIOR

Frontal section

FIGURE 4.5 | *Right hip (coxal) joint*

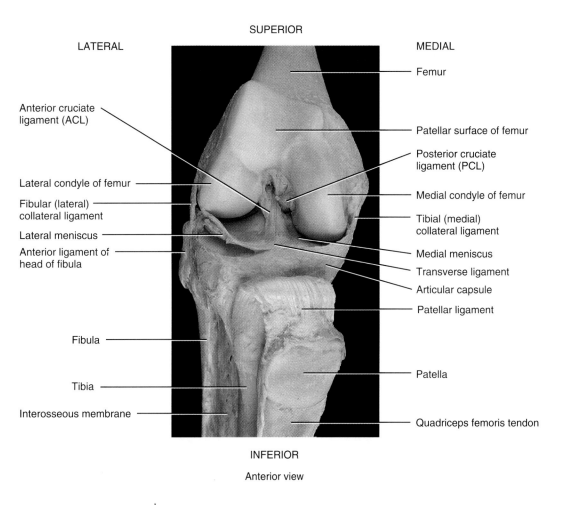

SUPERIOR

LATERAL

MEDIAL

Femur

Anterior cruciate
ligament (ACL)

Patellar surface of femur

Posterior cruciate
ligament (PCL)

Lateral condyle of femur

Medial condyle of femur

Fibular (lateral)
collateral ligament

Tibial (medial)
collateral ligament

Lateral meniscus

Medial meniscus

Anterior ligament of
head of fibula

Transverse ligament

Articular capsule

Patellar ligament

Fibula

Patella

Tibia

Interosseous membrane

Quadriceps femoris tendon

INFERIOR

Anterior view

FIGURE 4.6 | *Right knee, flexed*

SUPERIOR

Intercondylar fossa

Femur

Anterior cruciate ligament

Lateral condyle of femur (covered with articular cartilage)

Medial condyle of femur (covered with articular cartilage)

Fibular (lateral) collateral ligament

Lateral meniscus

Tibial (medial) collateral ligament

Posterior cruciate ligament

Medial meniscus

Posterior ligament of tibiofibular joint

Tibia

Fibula

Interosseus membrane

MEDIAL

LATERAL

INFERIOR

Posterior view

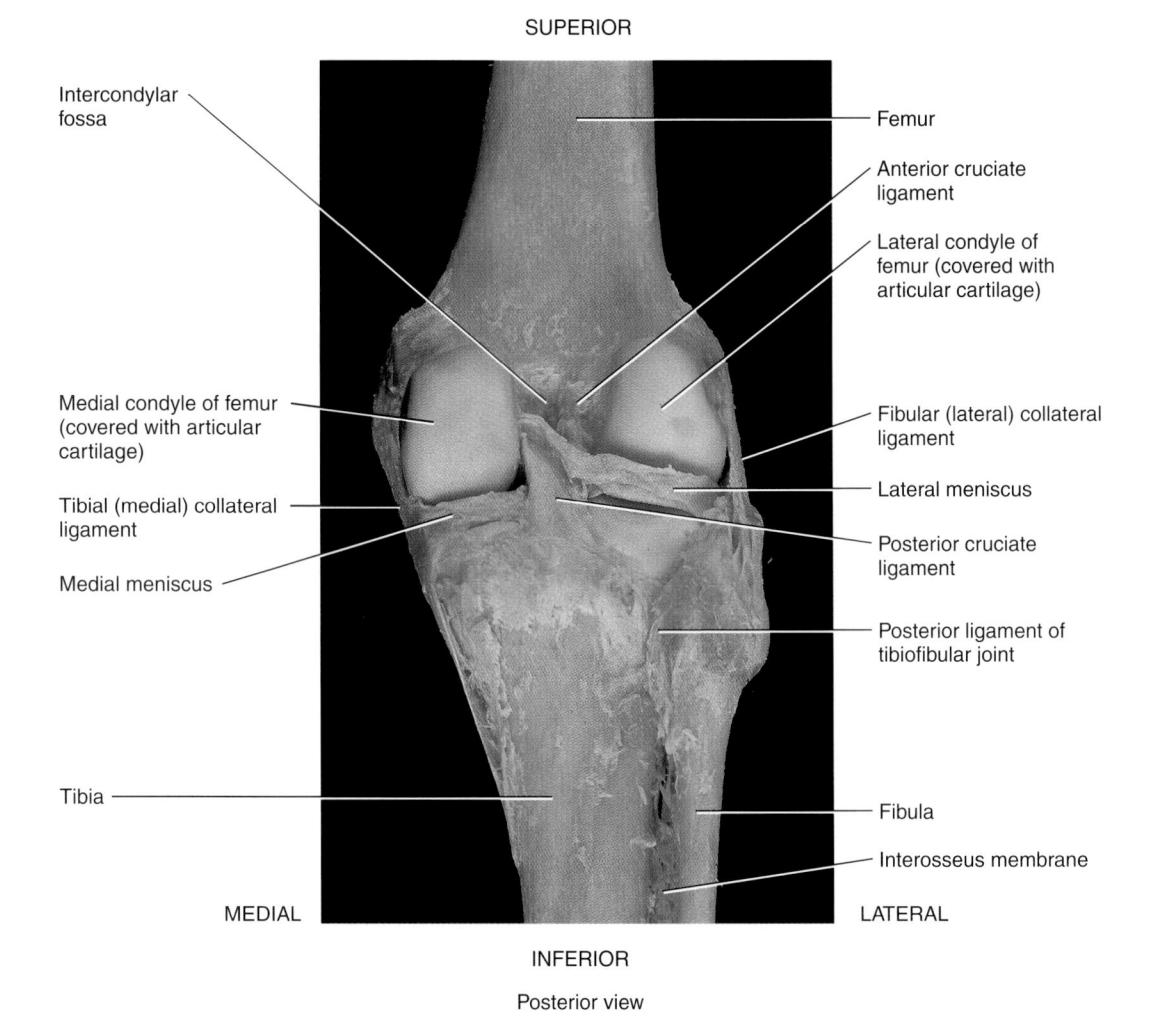

FIGURE 4.7 | *Right knee*

SUPERIOR

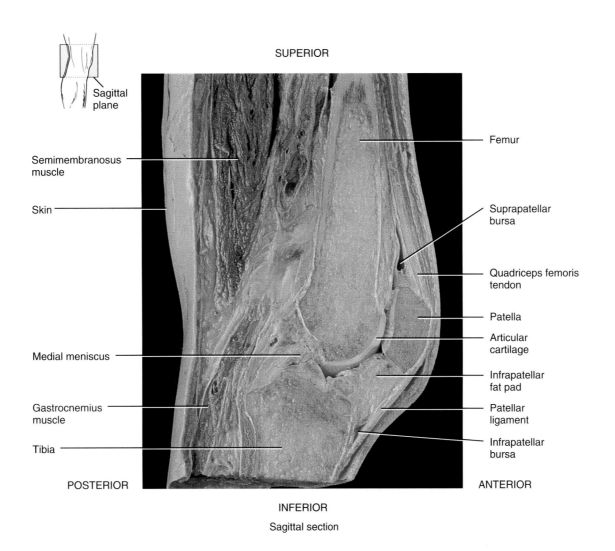

Sagittal
plane

Femur

Semimembranosus
muscle

Skin

Suprapatellar
bursa

Quadriceps femoris
tendon

Patella

Articular
cartilage

Medial meniscus

Infrapatellar
fat pad

Gastrocnemius
muscle

Patellar
ligament

Tibia

Infrapatellar
bursa

POSTERIOR

ANTERIOR

INFERIOR

Sagittal section

FIGURE 4.8 | *Right knee*

TABLE 5.1 | *Characteristics Used to Name Muscles*

### DIRECTION: Orientation of muscle relative to the body's midline

| Name | Meaning | Example |
|------|---------|---------|
| Rectus | parallel to midline | Rectus abdominis |
| Transverse | perpendicular to midline | Transversus abdominis |
| Oblique | diagonal to midline | External oblique |

### SIZE: Relative size of the muscle

| Name | Meaning | Example |
|------|---------|---------|
| Maximus | largest | Gluteus maximus |
| Minimus | smallest | Gluteus minimus |
| Longus | longest | Adductor longus |
| Brevis | shortest | Peroneus brevis |
| Latissimus | widest | Latissimus dorsi |
| Longissimus | longest | Longissimus capitis |
| Magnus | large | Adductor magnus |
| Major | larger | Pectoralis major |
| Minor | smaller | Pectoralis minor |
| Vastus | great | Vastus lateralis |

### SHAPE: Relative shape of the muscle

| Name | Meaning | Example |
|------|---------|---------|
| Deltoid | triangular | Deltoid |
| Trapezius | trapezoid | Trapezius |
| Serratus | saw-toothed | Serratus anterior |
| Rhomboideus | diamond-shaped | Romboideus major |
| Orbicularis | circular | Orbicularis oculi |
| Pectinate | comblike | Pectineus |
| Piriformis | pear-shaped | Piriformis |
| Platys | flat | Platysma |
| Quadratus | square | Quadratus femoris |
| Gracilis | slender | Gracilis |

### ACTION: Principal action of the muscle

| Name | Meaning | Example |
|------|---------|---------|
| Flexor | decreases joint angle | Flexor carpi radialis |
| Extensor | increases joint angle | Extensor carpi ulnaris |
| Abductor | moves bone away from midline | Abductor pollicis longus |
| Adductor | moves bone closer to midline | Adductor longus |
| Levator | produces superior movement | Levator scapulae |
| Depressor | produces inferior movement | Depressor labii inferioris |
| Supinator | turns palm superiorly or anteriorly | Supinator |
| Pronator | turns palm inferiorly or posteriorly | Pronator teres |
| Sphincter | decreases size of opening | External anal sphincter |
| Tensor | makes a body part rigid | Tensor fasciae latae |
| Rotator | moves bone around longitudinal axis | Obturator externus |

### LOCATION: Structure near which a muscle is found

**Example**

Frontalis, a muscle near the frontal bone; tibialis anterior, a muscle near the front of the tibia.

### ORIGIN AND INSERTION: Sites where muscle originates and inserts

**Example**

Sternocleidomastoid, originates on the sternum and clavicle and inserts on mastoid process of temporal bone; stylohyoid, originates on styloid process of temporal bone and inserts on the hyoid bone.

### NUMBER OF ORIGINS: Number of tendons of origin

| Name | Meaning | Example |
|------|---------|---------|
| Biceps | Two origins | Biceps brachii |
| Triceps | Three origins | Triceps brachii |
| Quadriceps | Four origins | Quadriceps femoris |

FIGURE 5.1

**Principal superficial skeletal muscles**

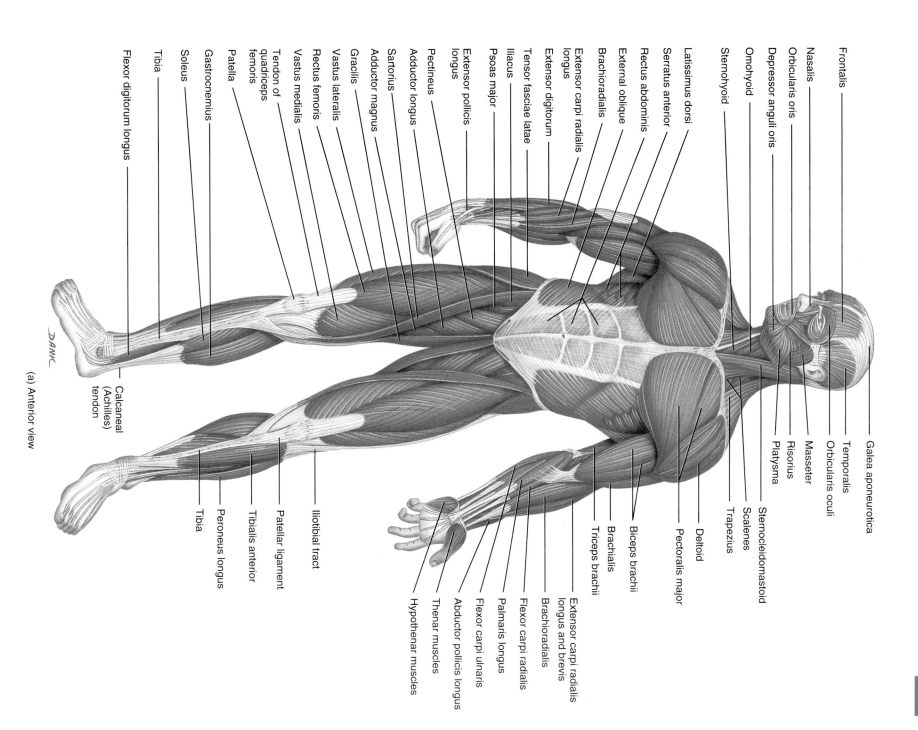

(a) Anterior view

Galea aponeurotica
Frontalis
Temporalis
Orbicularis oculi
Nasalis
Orbicularis oris
Masseter
Depressor anguli oris
Risorius
Omohyoid
Platysma
Sternohyoid
Sternocleidomastoid
Scalenes
Trapezius
Latissimus dorsi
Serratus anterior
Rectus abdominis
External oblique
Brachioradialis
Extensor carpi radialis longus
Extensor pollicis longus
Extensor digitorum
Tensor fasciae latae
Psoas major
Iliacus
Pectineus
Sartorius
Adductor longus
Adductor magnus
Gracilis
Vastus lateralis
Rectus femoris
Vastus medialis
Tendon of quadriceps femoris
Patella
Gastrocnemius
Soleus
Tibia
Flexor digitorum longus

Deltoid
Pectoralis major
Biceps brachii
Triceps brachii
Brachialis
Extensor carpi radialis longus and brevis
Brachioradialis
Flexor carpi radialis
Palmaris longus
Flexor carpi ulnaris
Abductor pollicis longus
Thenar muscles
Hypothenar muscles
Iliotibial tract
Patellar ligament
Tibialis anterior
Peroneus longus
Tibia
Calcaneal (Achilles) tendon

65

*Principal superficial skeletal muscles*

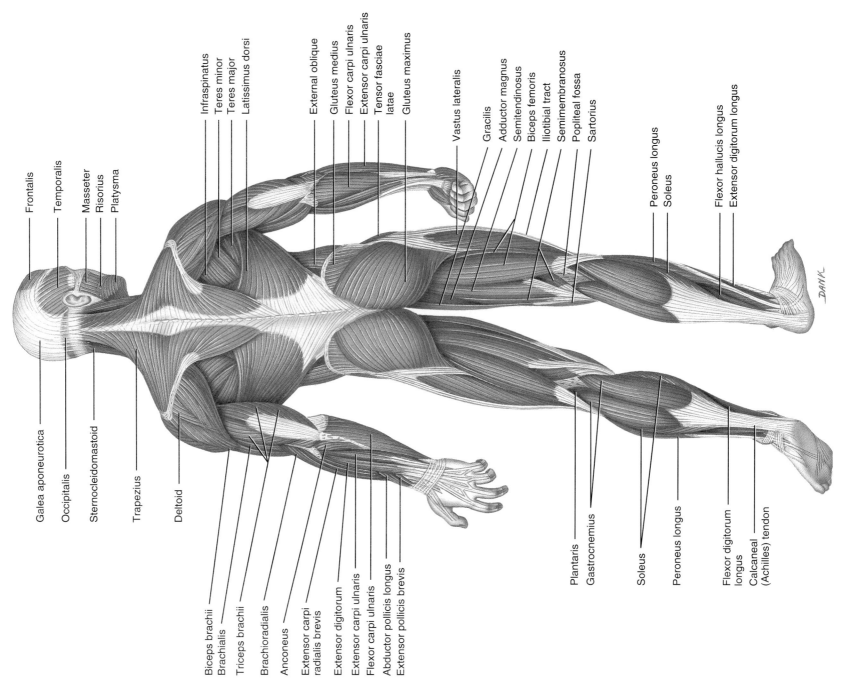

Frontalis
Temporalis
Masseter
Risorius
Platysma

Galea aponeurotica
Occipitalis
Sternocleidomastoid
Trapezius
Deltoid

Infraspinatus
Teres minor
Teres major
Latissimus dorsi

External oblique
Gluteus medius
Flexor carpi ulnaris
Extensor carpi ulnaris
Tensor fasciae latae
Gluteus maximus

Vastus lateralis
Gracilis
Adductor magnus
Semitendinosus
Biceps femoris
Iliotibial tract
Semimembranosus
Popliteal fossa
Sartorius

Peroneus longus
Soleus

Flexor hallucis longus
Extensor digitorum longus

Biceps brachii
Brachialis
Triceps brachii
Brachioradialis
Anconeus
Extensor carpi radialis brevis
Extensor digitorum
Extensor carpi ulnaris
Flexor carpi ulnaris
Abductor pollicis longus
Extensor pollicis brevis

Plantaris
Gastrocnemius
Soleus
Peroneus longus
Flexor digitorum longus
Calcaneal (Achilles) tendon

DANK

(b) Posterior view

FIGURE 5.2 | *Muscles of facial expression*

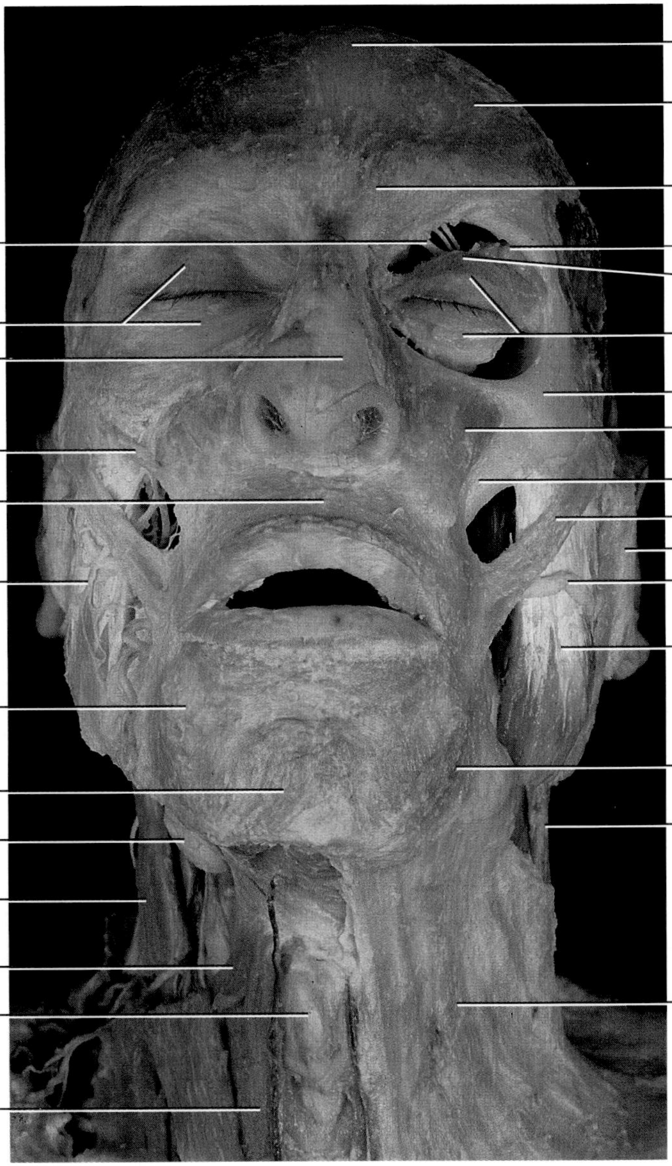

Frontal bone

Frontalis

Corrugator supercilii

Supraorbital nerve

Lacrimal gland

Levator palpebrae superioris

Orbicularis oculi

Tarsal plates

Nasal cartilage

Zygomatic bone

Levator labii superioris

Zygomaticus minor

Maxilla

Orbicularis oris

Zygomaticus major

Parotid gland

Facial (VII) nerve

Parotid gland duct

Masseter

Depressor labii inferioris

Depressor anguli oris

Mentalis

External jugular vein

Submandibular gland

Sternocleidomastoid (cut)

Omohyoid

Thyroid cartilage (Adam's apple)

Platysma

Sternohyoid

INFERIOR

Anterior view

SUPERIOR

POSTERIOR

ANTERIOR

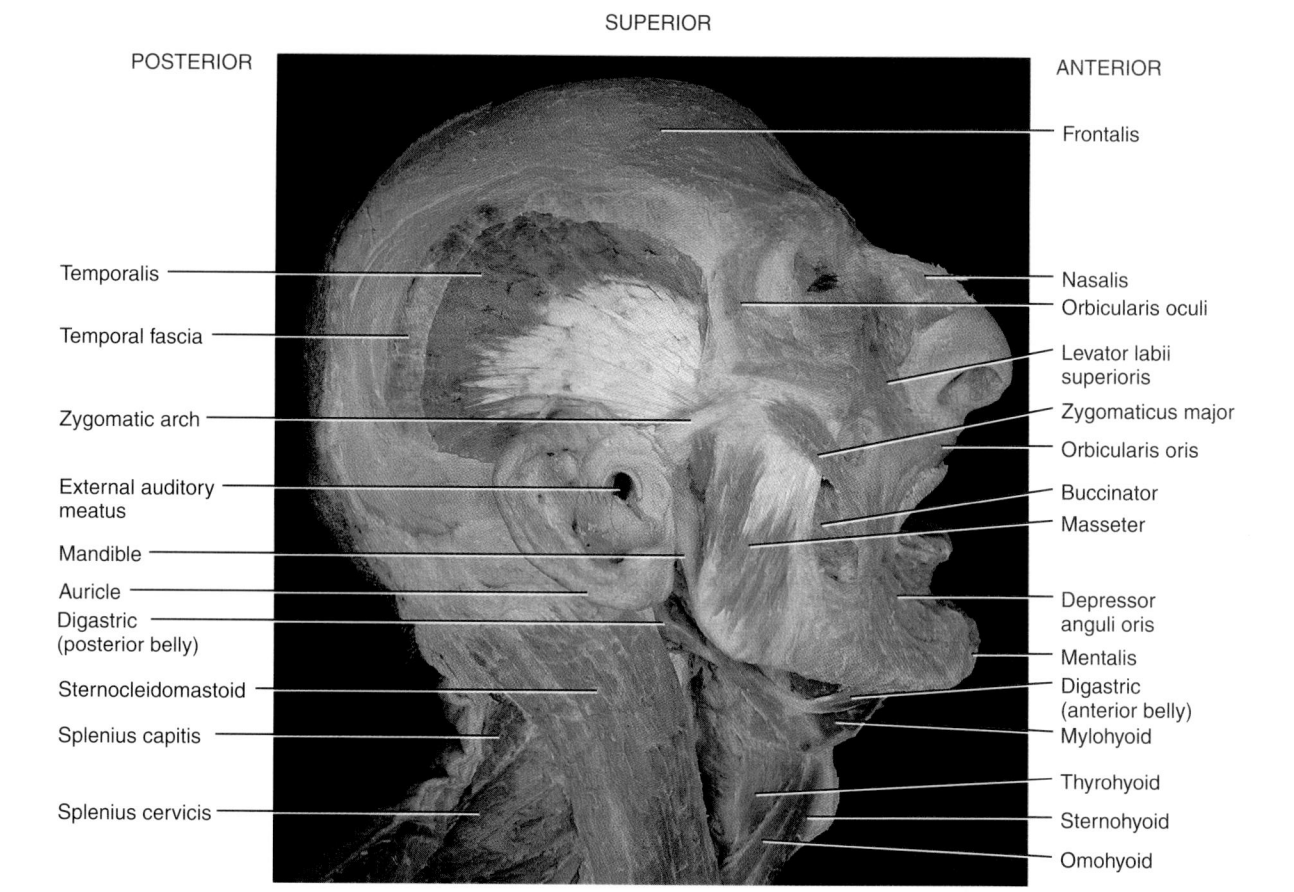

Frontalis

Temporalis

Nasalis
Orbicularis oculi

Temporal fascia

Levator labii
superioris

Zygomatic arch

Zygomaticus major

Orbicularis oris

External auditory
meatus

Buccinator

Masseter

Mandible

Auricle

Digastric
(posterior belly)

Depressor
anguli oris

Sternocleidomastoid

Mentalis

Digastric
(anterior belly)

Splenius capitis

Mylohyoid

Thyrohyoid

Splenius cervicis

Sternohyoid

Omohyoid

INFERIOR

Right lateral view

FIGURE 5.3 | *Muscles of facial
expression*

SUPERIOR

POSTERIOR

ANTERIOR

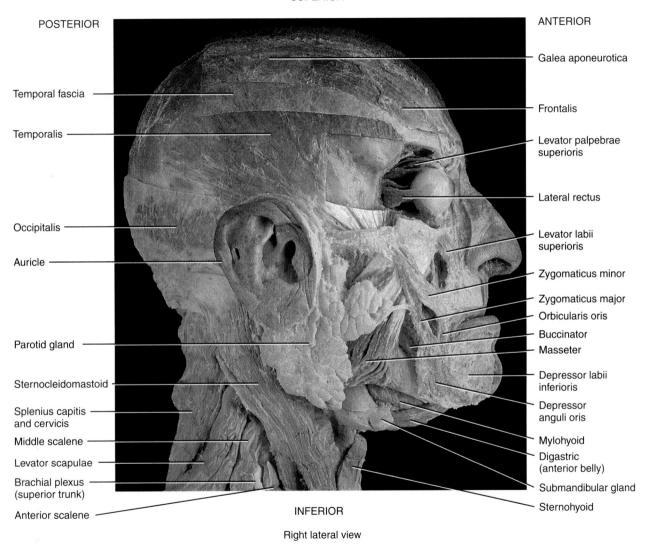

Galea aponeurotica

Temporal fascia

Frontalis

Temporalis

Levator palpebrae
superioris

Lateral rectus

Occipitalis

Levator labii
superioris

Auricle

Zygomaticus minor

Zygomaticus major

Orbicularis oris

Buccinator

Parotid gland

Masseter

Depressor labii
inferioris

Sternocleidomastoid

Depressor
anguli oris

Splenius capitis
and cervicis

Middle scalene

Mylohyoid

Levator scapulae

Digastric
(anterior belly)

Brachial plexus
(superior trunk)

Submandibular gland

Anterior scalene

Sternohyoid

INFERIOR

Right lateral view

FIGURE 5.4 | *Muscles of facial
expression*

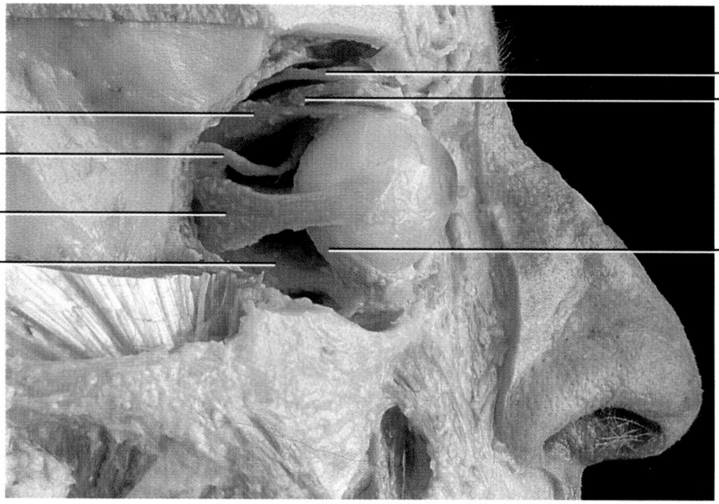

SUPERIOR

Superior oblique

Levator palpebrae superioris

Superior rectus

Medial rectus

Lateral rectus

Inferior oblique

Inferior rectus

POSTERIOR

ANTERIOR

INFERIOR

Lateral view

FIGURE 5.5 | *Extrinsic muscles of the right eyeball*

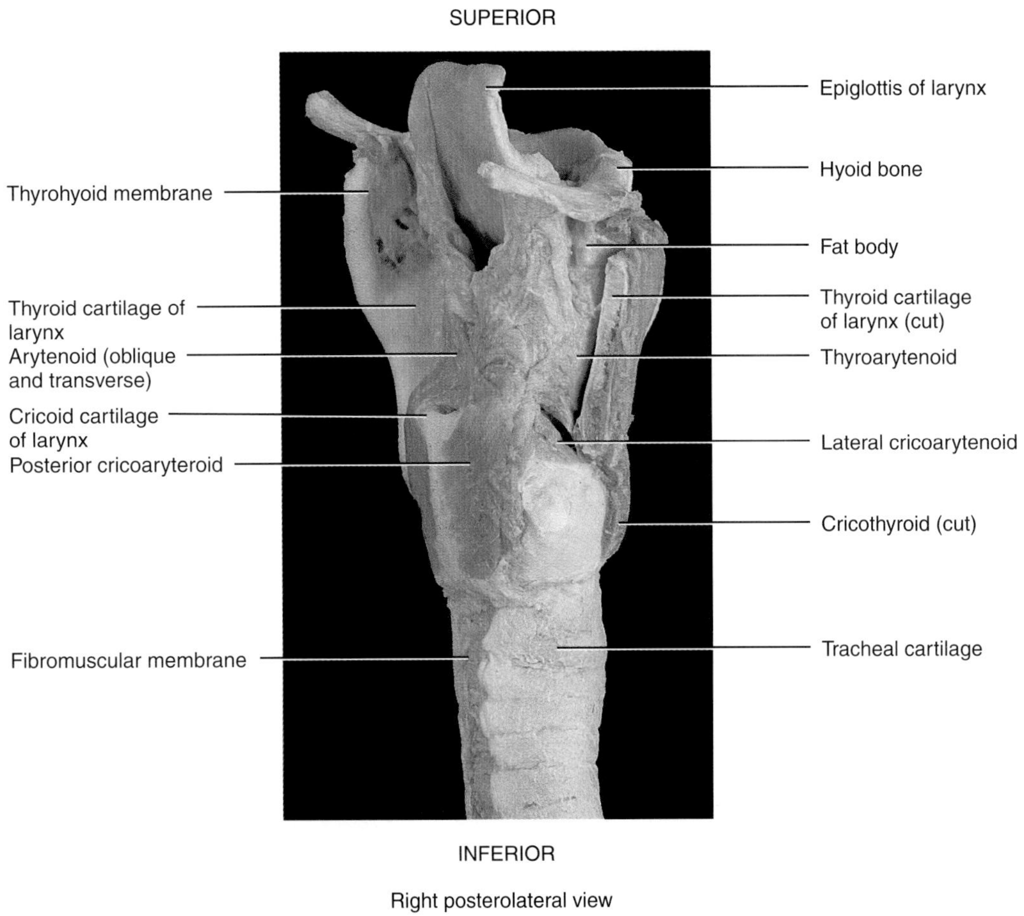

SUPERIOR

Epiglottis of larynx

Hyoid bone

Thyrohyoid membrane

Fat body

Thyroid cartilage of
larynx

Thyroid cartilage
of larynx (cut)

Arytenoid (oblique
and transverse)

Thyroarytenoid

Cricoid cartilage
of larynx

Posterior cricoaryteroid

Lateral cricoarytenoid

Cricothyroid (cut)

Fibromuscular membrane

Tracheal cartilage

INFERIOR

Right posterolateral view

FIGURE 5.6 | *Muscles of the larynx*

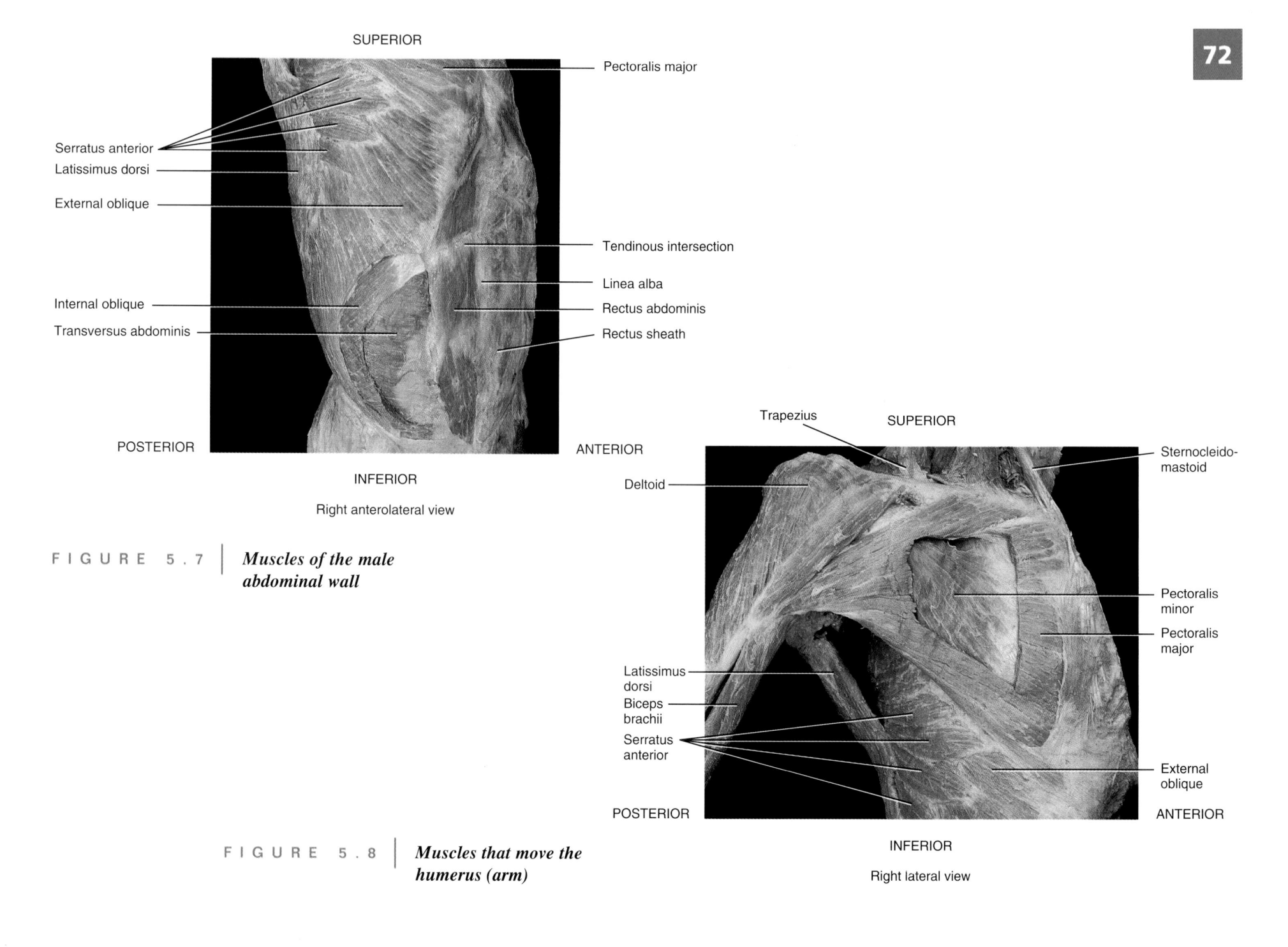

SUPERIOR

Pectoralis major

Serratus anterior

Latissimus dorsi

External oblique

Tendinous intersection

Linea alba

Internal oblique

Rectus abdominis

Transversus abdominis

Rectus sheath

POSTERIOR

ANTERIOR

INFERIOR

Right anterolateral view

FIGURE 5.7 | *Muscles of the male abdominal wall*

Trapezius

SUPERIOR

Deltoid

Sternocleido-mastoid

Pectoralis minor

Pectoralis major

Latissimus dorsi

Biceps brachii

Serratus anterior

External oblique

POSTERIOR

ANTERIOR

INFERIOR

FIGURE 5.8 | *Muscles that move the humerus (arm)*

Right lateral view

SUPERIOR

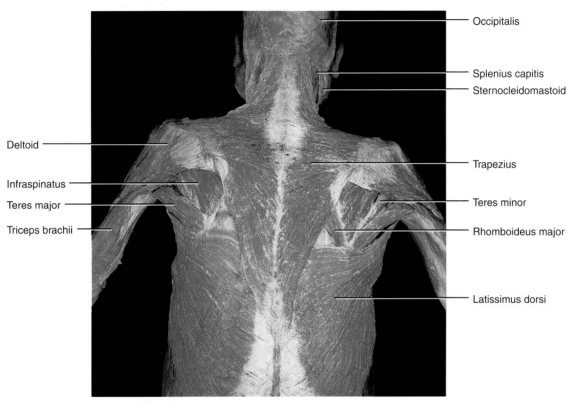

Occipitalis

Splenius capitis
Sternocleidomastoid

Deltoid

Trapezius

Infraspinatus

Teres major

Teres minor

Triceps brachii

Rhomboideus major

Latissimus dorsi

INFERIOR

Posterior view

FIGURE 5.9 | *Muscles that move the humerus (arm)*

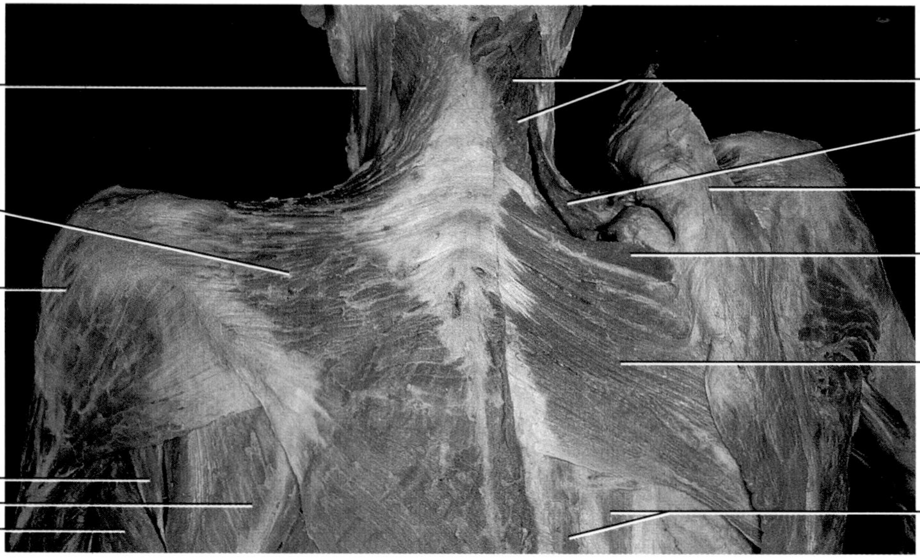

SUPERIOR

Sternocleidomastoid

Trapezius

Deltoid

Teres minor
Infraspinatus
Teres major

Splenius capitis
and cervicis

Levator scapulae

Trapezius (reflected)

Rhomboideus minor

Rhomboideus major

Erector spinae

INFERIOR

Posterior view

FIGURE  5 . 1 0  | *Muscles that move
the humerus (arm)*

SUPERIOR

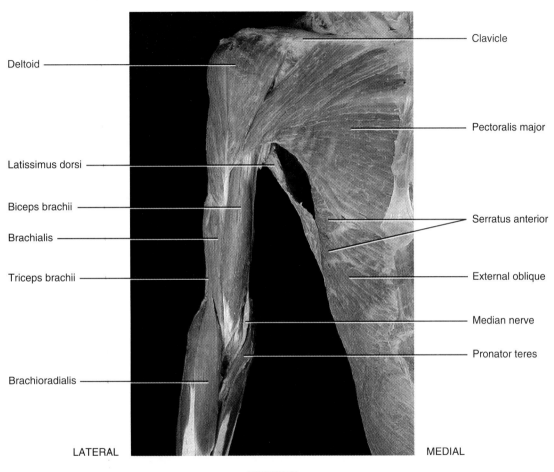

Deltoid —————————————————

Clavicle

Latissimus dorsi —————————

Pectoralis major

Biceps brachii —————————

Brachialis —————————————

Serratus anterior

Triceps brachii —————————

External oblique

Median nerve

Pronator teres

Brachioradialis —————————

LATERAL                                    MEDIAL

INFERIOR

Anterior view

FIGURE  5.1-1 | *Muscles that move the right radius and ulna (forearm)*

75

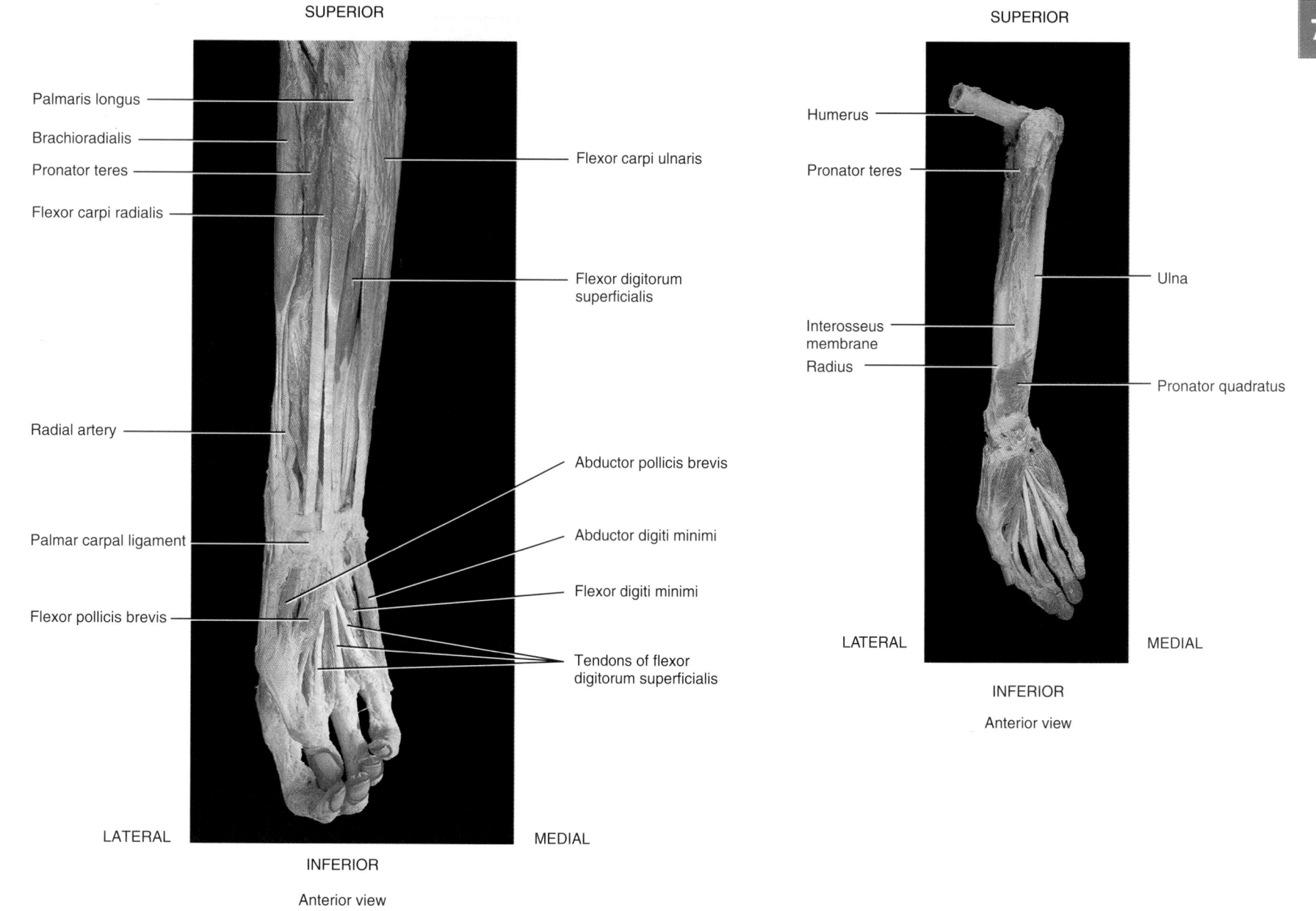

SUPERIOR

Palmaris longus

Brachioradialis

Pronator teres

Flexor carpi radialis

Radial artery

Palmar carpal ligament

Flexor pollicis brevis

Flexor carpi ulnaris

Flexor digitorum superficialis

Abductor pollicis brevis

Abductor digiti minimi

Flexor digiti minimi

Tendons of flexor digitorum superficialis

LATERAL

MEDIAL

INFERIOR

Anterior view

SUPERIOR

Humerus

Pronator teres

Interosseus membrane

Radius

Ulna

Pronator quadratus

LATERAL

MEDIAL

INFERIOR

Anterior view

FIGURE 5.12 | *Muscles that move the right wrist, hand, and fingers*

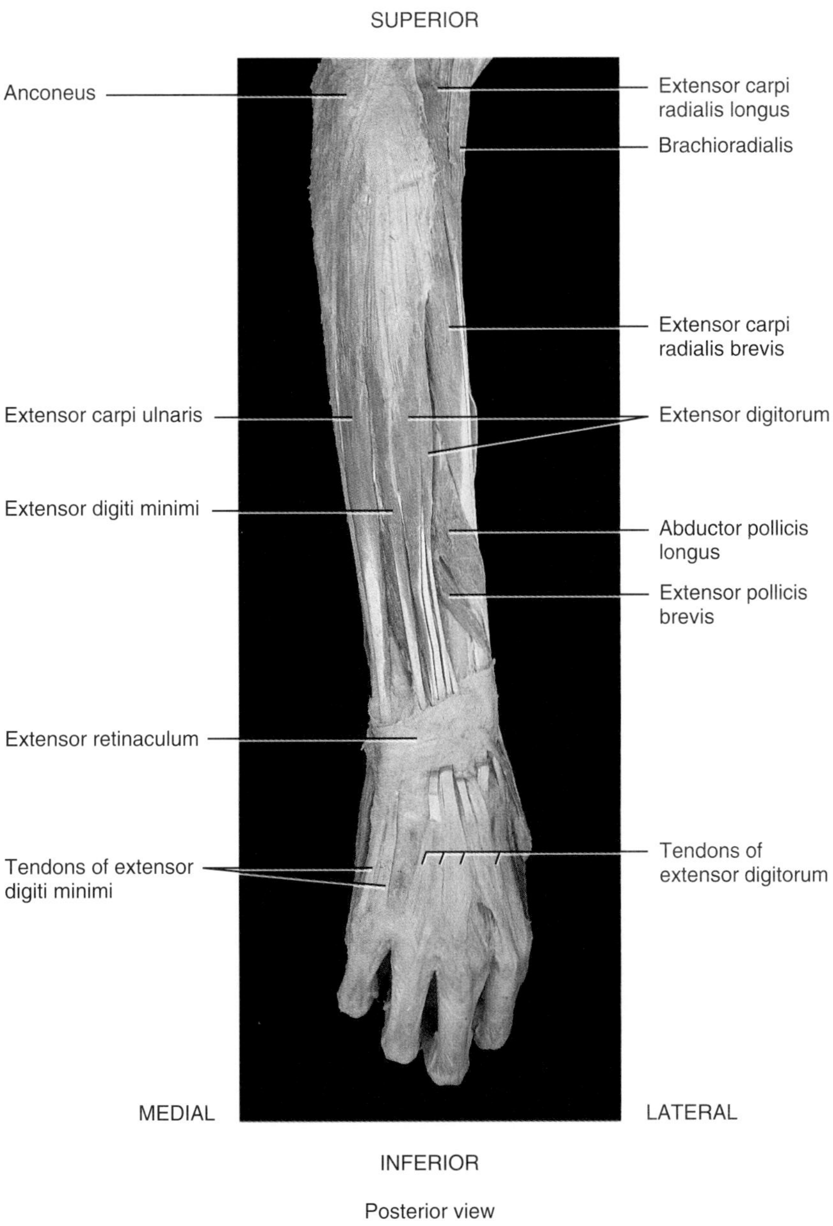

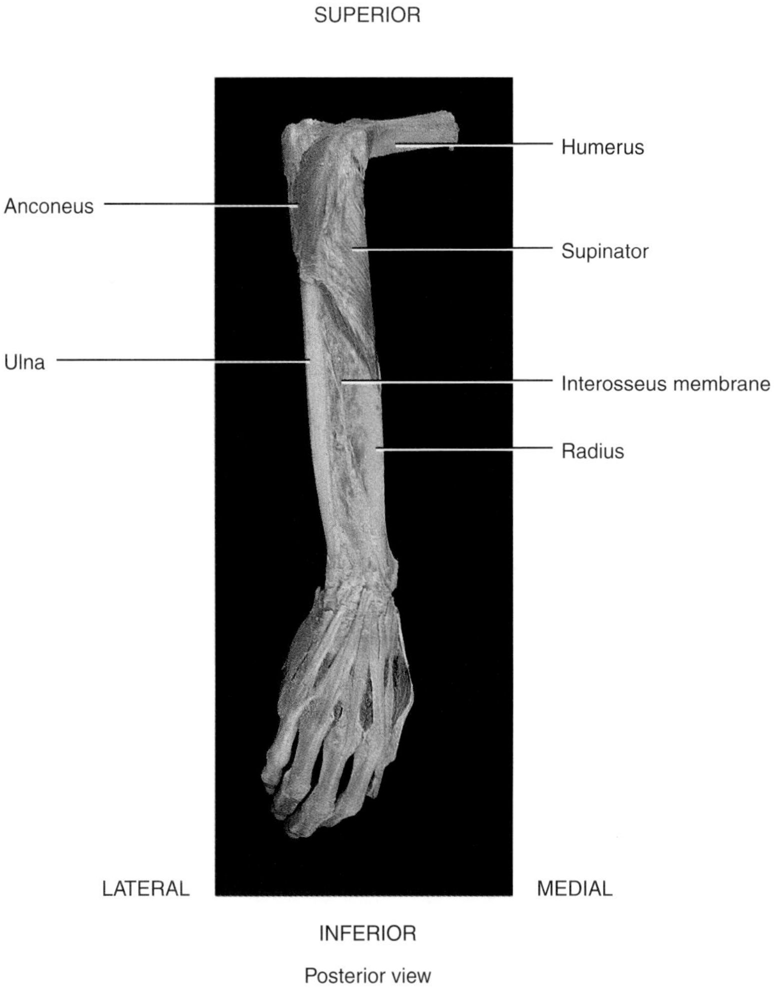

SUPERIOR

Anconeus

Extensor carpi
radialis longus

Brachioradialis

Extensor carpi
radialis brevis

Extensor carpi ulnaris

Extensor digitorum

Extensor digiti minimi

Abductor pollicis
longus

Extensor pollicis
brevis

Extensor retinaculum

Tendons of extensor
digiti minimi

Tendons of
extensor digitorum

MEDIAL

LATERAL

INFERIOR

Posterior view

SUPERIOR

Anconeus

Humerus

Supinator

Ulna

Interosseus membrane

Radius

LATERAL

MEDIAL

INFERIOR

Posterior view

FIGURE 5.13 | *Muscles that move the right wrist, hand, and fingers*

77

SUPERIOR

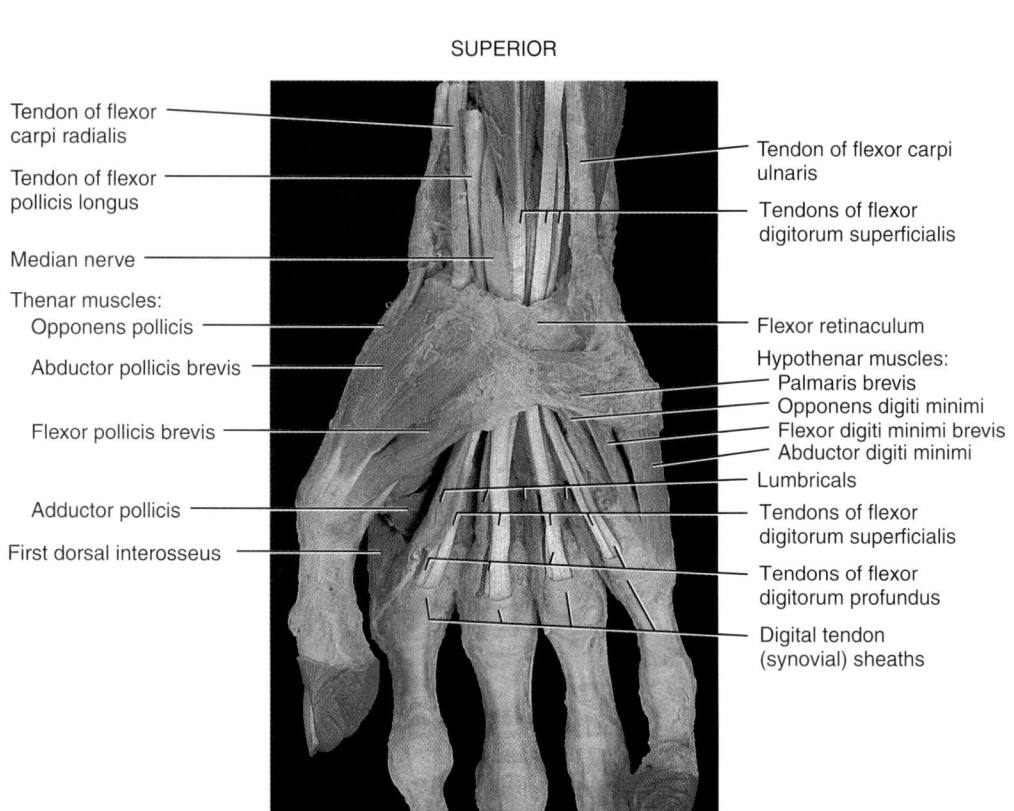

Tendon of flexor
carpi radialis

Tendon of flexor
pollicis longus

Median nerve

Thenar muscles:
Opponens pollicis

Abductor pollicis brevis

Flexor pollicis brevis

Adductor pollicis

First dorsal interosseus

Tendon of flexor carpi
ulnaris

Tendons of flexor
digitorum superficialis

Flexor retinaculum

Hypothenar muscles:
Palmaris brevis
Opponens digiti minimi
Flexor digiti minimi brevis
Abductor digiti minimi

Lumbricals

Tendons of flexor
digitorum superficialis

Tendons of flexor
digitorum profundus

Digital tendon
(synovial) sheaths

LATERAL

MEDIAL

INFERIOR

Anterior view

FIGURE 5.14 | *Intrinsic muscles of
the right hand*

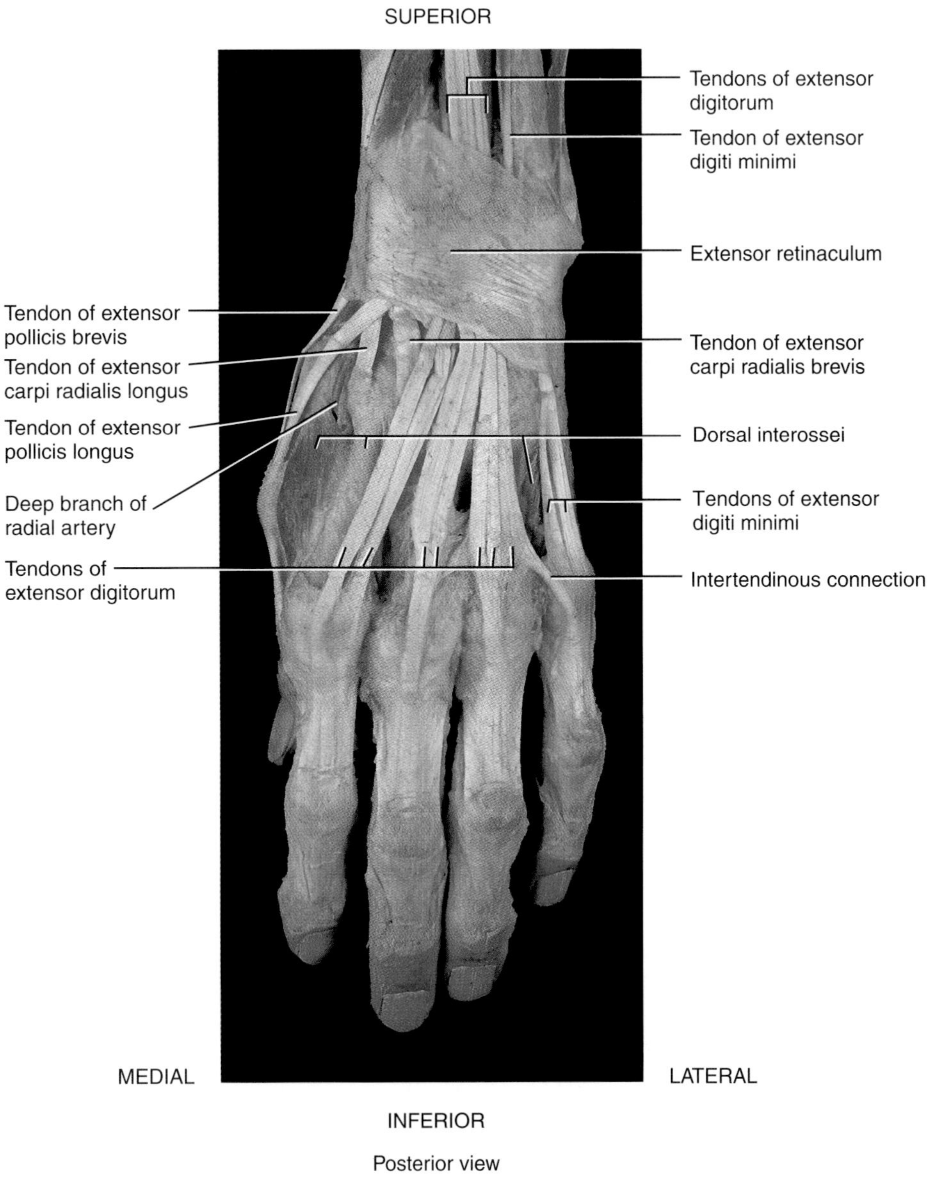

SUPERIOR

Tendons of extensor digitorum

Tendon of extensor digiti minimi

Extensor retinaculum

Tendon of extensor pollicis brevis

Tendon of extensor carpi radialis longus

Tendon of extensor pollicis longus

Deep branch of radial artery

Tendons of extensor digitorum

Tendon of extensor carpi radialis brevis

Dorsal interossei

Tendons of extensor digiti minimi

Intertendinous connection

MEDIAL

LATERAL

INFERIOR

Posterior view

FIGURE 5.15 | *Intrinsic muscles of the left hand*

79

SUPERIOR

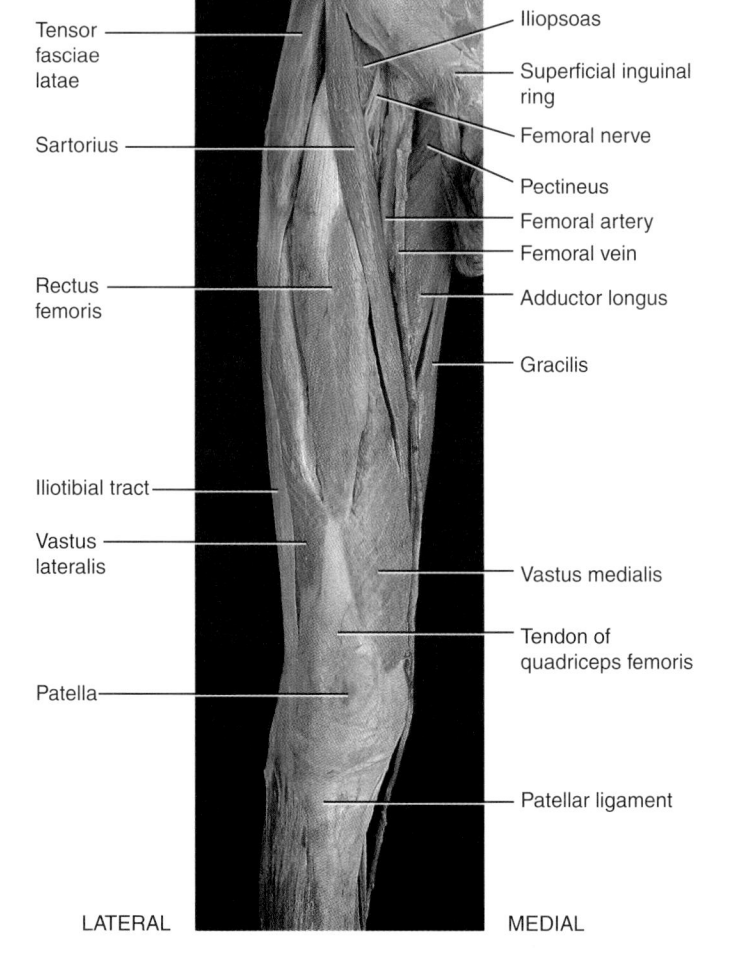

Rectus sheath

Tensor fasciae latae

Iliopsoas

Superficial inguinal ring

Sartorius

Femoral nerve

Pectineus

Femoral artery

Femoral vein

Rectus femoris

Adductor longus

Gracilis

Iliotibial tract

Vastus lateralis

Vastus medialis

Tendon of quadriceps femoris

Patella

Patellar ligament

LATERAL

MEDIAL

INFERIOR

Anterior view

FIGURE 5.16 | *Muscles that move the right femur (thigh) and tibia and fibula (leg)*

SUPERIOR

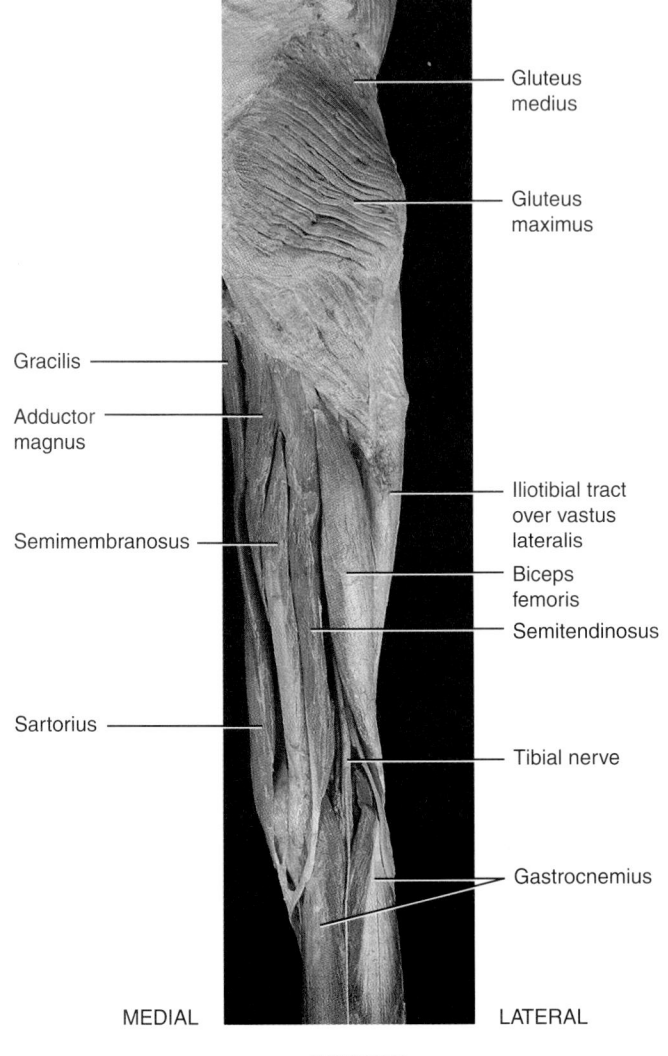

Gluteus medius

Gluteus maximus

Gracilis

Adductor magnus

Iliotibial tract over vastus lateralis

Semimembranosus

Biceps femoris

Semitendinosus

Sartorius

Tibial nerve

Gastrocnemius

MEDIAL

LATERAL

INFERIOR

Posterior view

FIGURE 5.17 | *Muscles that move the right femur (thigh) and tibia and fibula (leg)*

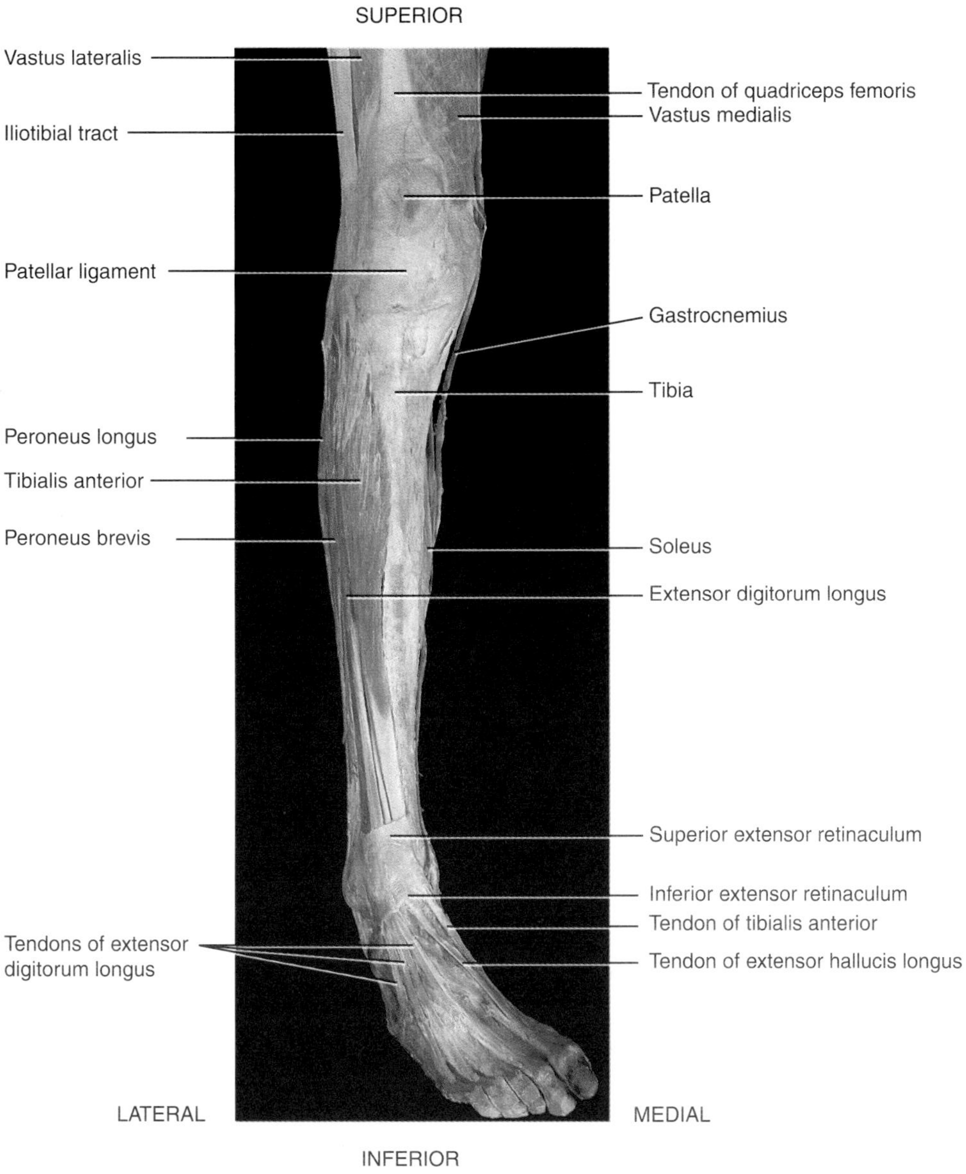

SUPERIOR

Vastus lateralis

Iliotibial tract

Patellar ligament

Peroneus longus

Tibialis anterior

Peroneus brevis

Tendon of quadriceps femoris
Vastus medialis

Patella

Gastrocnemius

Tibia

Soleus

Extensor digitorum longus

Superior extensor retinaculum

Inferior extensor retinaculum
Tendon of tibialis anterior
Tendon of extensor hallucis longus

Tendons of extensor
digitorum longus

LATERAL

MEDIAL

INFERIOR

Anterior view

FIGURE 5.18 | *Muscles that move the right foot and toes*

SUPERIOR

Biceps femoris

Vastus lateralis

Iliotibial tract (cut)

Plantaris

Soleus

Peroneus longus

Peroneus brevis

Semitendinosus

Semimembranosus

Gracilis

Tibial nerve

Gastrocnemius

Calcaneal
(Achilles) tendon

LATERAL

MEDIAL

INFERIOR

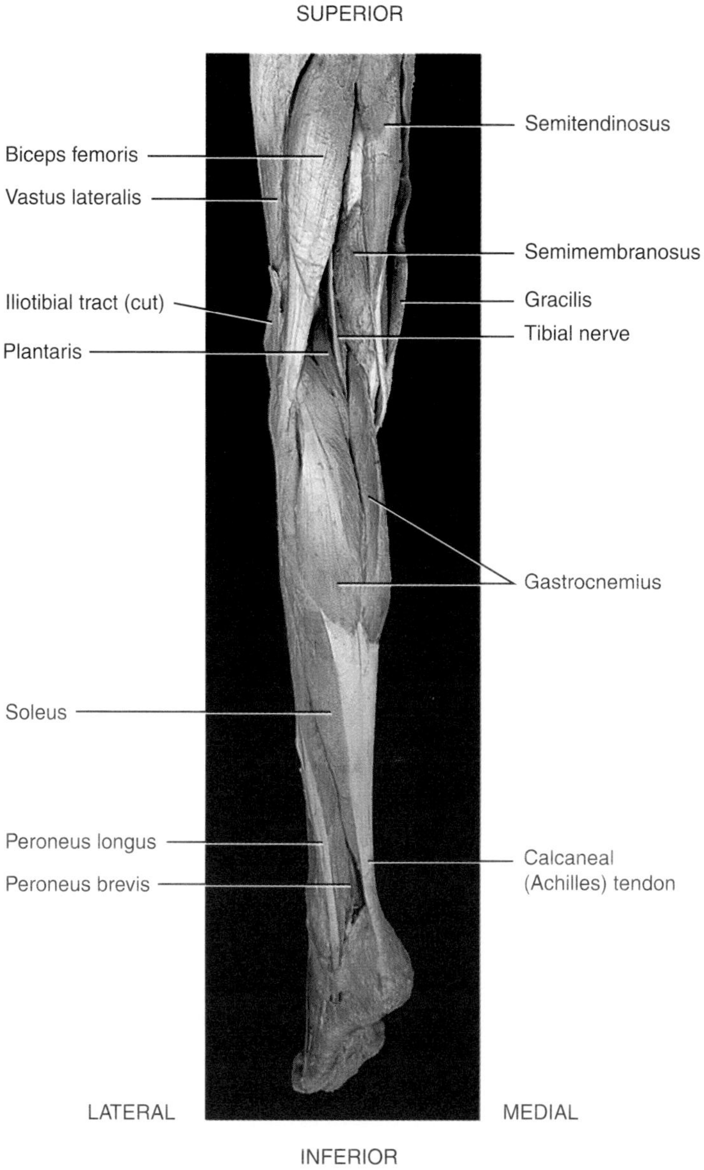

Posterior view

FIGURE 5.19 | *Muscles that move the right foot and toes*

# UNIT SIX | *The Cardiovascular System*

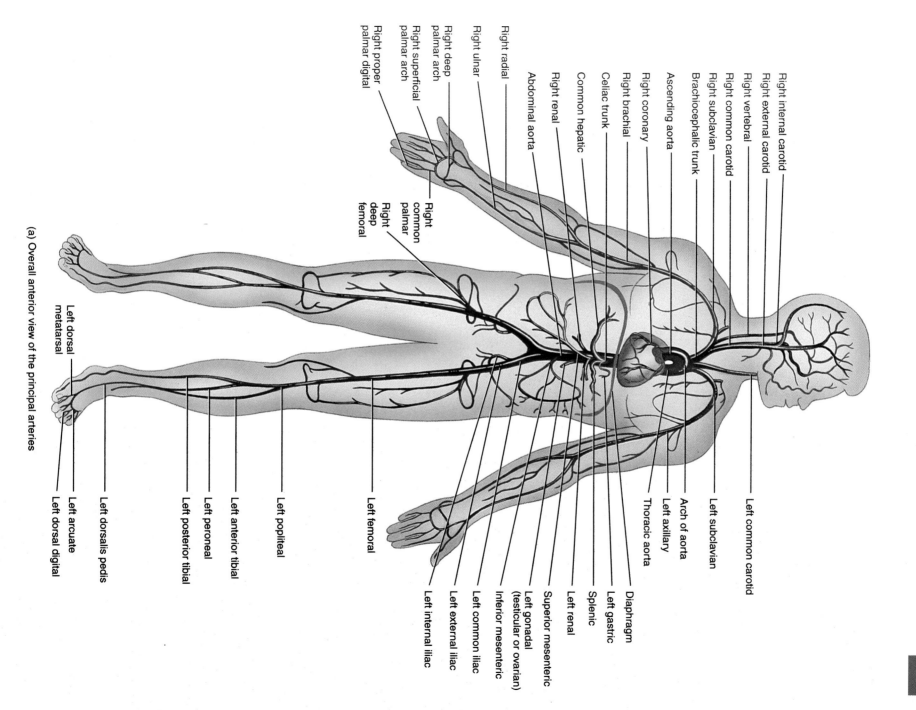

Right internal carotid

Right external carotid

Right vertebral

Right common carotid

Right subclavian

Brachiocephalic trunk

Ascending aorta

Right coronary

Right brachial

Celiac trunk

Common hepatic

Right renal

Abdominal aorta

Right ulnar

Right radial

Right deep palmar arch

Right superficial palmar arch

Right common palmar

Right deep femoral

Right proper palmar digital

Left common carotid

Left subclavian

Arch of aorta

Left axillary

Thoracic aorta

Diaphragm

Left gastric

Splenic

Left renal

Superior mesenteric

Left gonadal (testicular or ovarian)

Inferior mesenteric

Left common iliac

Left external iliac

Left internal iliac

Left femoral

Left popliteal

Left anterior tibial

Left peroneal

Left posterior tibial

Left dorsalis pedis

Left arcuate

Left dorsal metatarsal

Left dorsal digital

(a) Overall anterior view of the principal arteries

83

F I G U R E  6 . 1  | *Principal blood vessels*

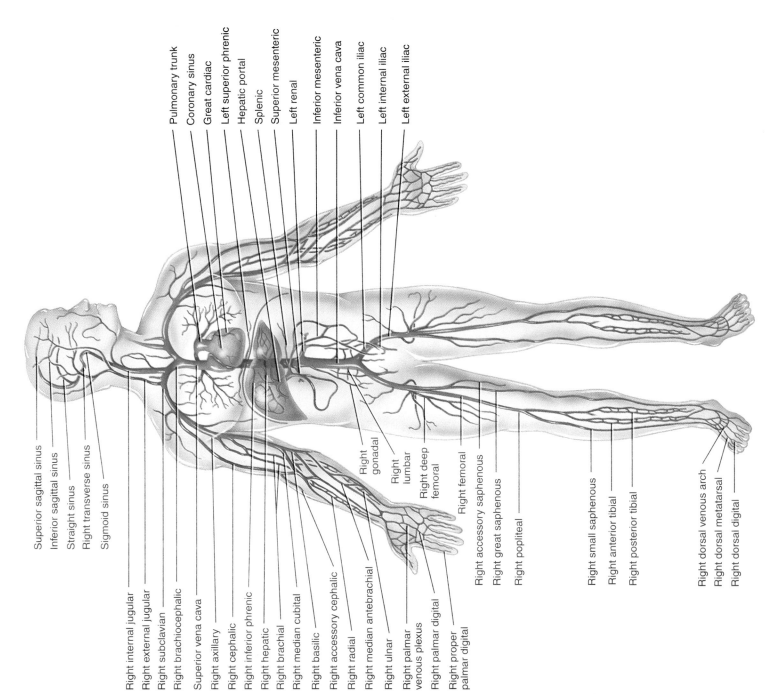

Superior sagittal sinus
Inferior sagittal sinus
Straight sinus
Right transverse sinus
Sigmoid sinus

Pulmonary trunk
Coronary sinus
Great cardiac
Left superior phrenic
Hepatic portal
Splenic
Superior mesenteric
Left renal
Inferior mesenteric
Inferior vena cava
Left common iliac
Left internal iliac
Left external iliac

Right internal jugular
Right external jugular
Right subclavian
Right brachiocephalic
Superior vena cava
Right axillary
Right cephalic
Right inferior phrenic
Right hepatic
Right brachial
Right median cubital
Right basilic
Right accessory cephalic
Right radial
Right median antebrachial
Right ulnar
Right palmar venous plexus
Right palmar digital
Right proper palmar digital

Right gonadal
Right lumbar
Right deep femoral

Right femoral
Right accessory saphenous
Right great saphenous
Right popliteal

Right small saphenous
Right anterior tibial
Right posterior tibial

Right dorsal venous arch
Right dorsal metatarsal
Right dorsal digital

(b) Overall anterior view of the principal veins

FIGURE 6.2 | *Histology of blood vessels*

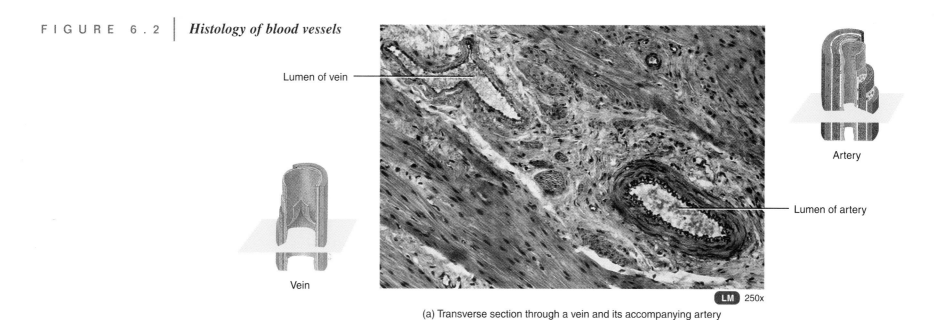

Lumen of vein

Vein

Artery

Lumen of artery

LM 250x

(a) Transverse section through a vein and its accompanying artery

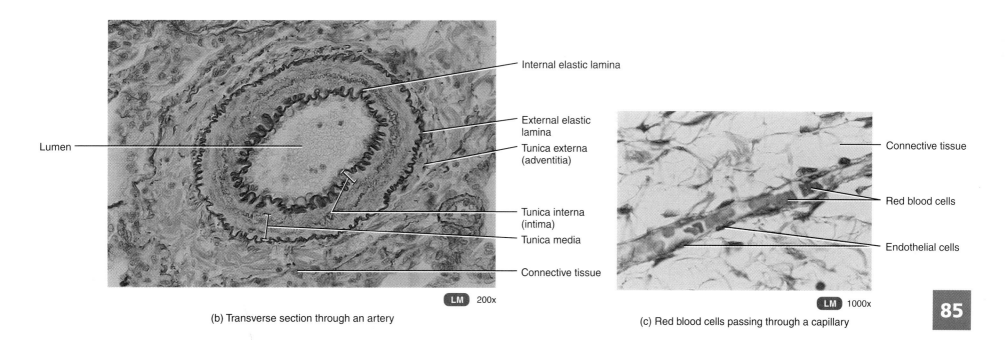

Lumen

Internal elastic lamina

External elastic lamina

Tunica externa (adventitia)

Tunica interna (intima)

Tunica media

Connective tissue

LM 200x

(b) Transverse section through an artery

Connective tissue

Red blood cells

Endothelial cells

LM 1000x

(c) Red blood cells passing through a capillary

85

SUPERIOR

Brachiocephalic vein

Brachiocephalic trunk

Superior vena cava

Ascending aorta

Right auricle of
right atrium

Adipose tissue over
right ventricle

Right ventricle

Left subclavian artery

Left common carotid artery

Arch of aorta

Ligamentum arteriosum

Left pulmonary artery

Pulmonary trunk

Left auricle of left atrium

Anterior interventricular
sulcus and vessels

Left ventricle

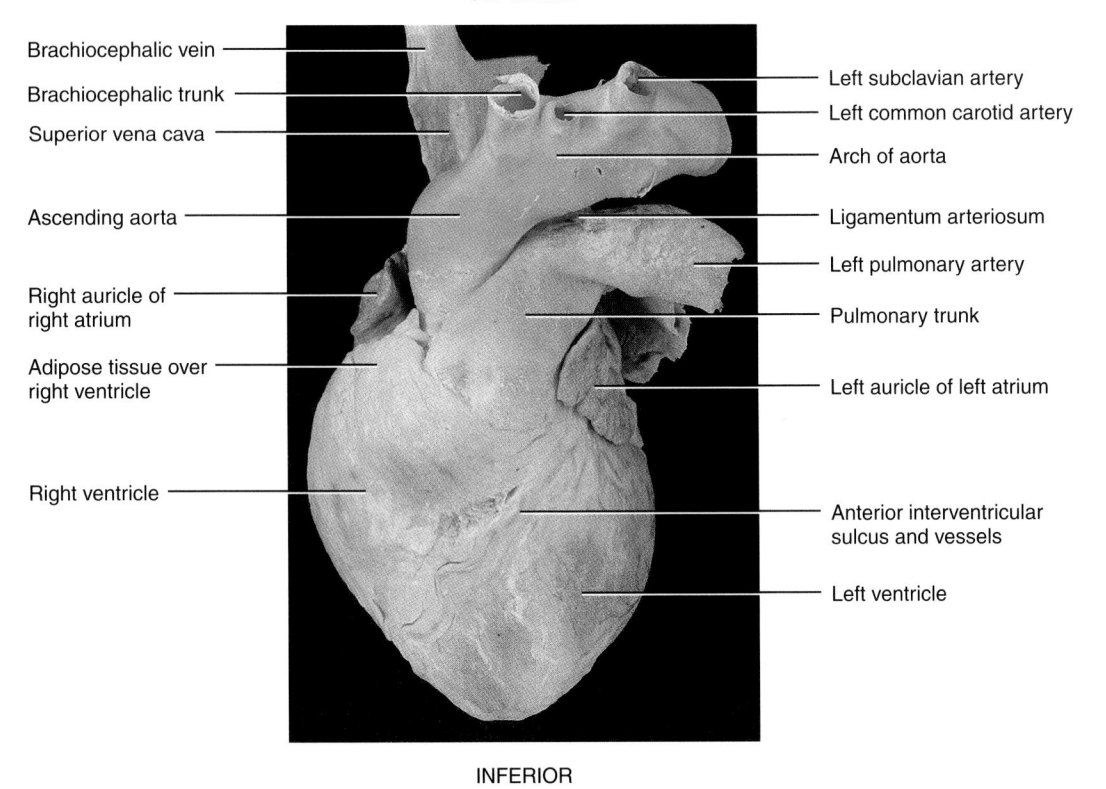

INFERIOR

(a) Anterior view of human heart

FIGURE 6.3 | *Heart*

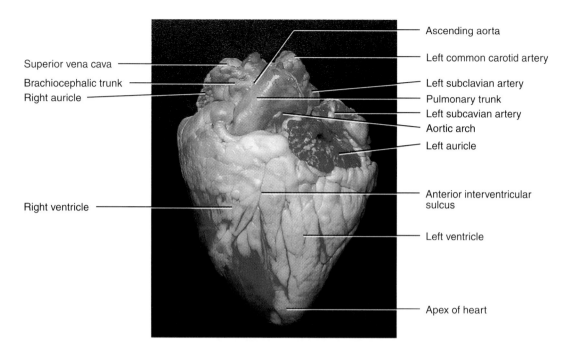

Superior vena cava

Brachiocephalic trunk

Right auricle

Right ventricle

Ascending aorta

Left common carotid artery

Left subclavian artery

Pulmonary trunk

Left subcavian artery

Aortic arch

Left auricle

Anterior interventricular sulcus

Left ventricle

Apex of heart

(b) Anterior view of sheep heart

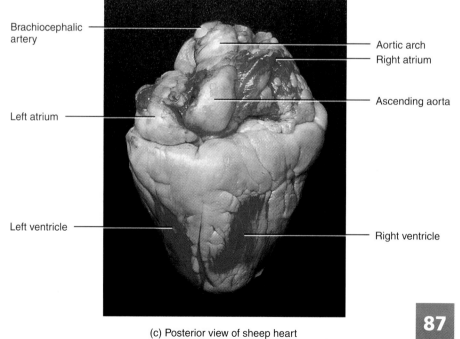

Brachiocephalic artery

Left atrium

Left ventricle

Aortic arch

Right atrium

Ascending aorta

Right ventricle

(c) Posterior view of sheep heart

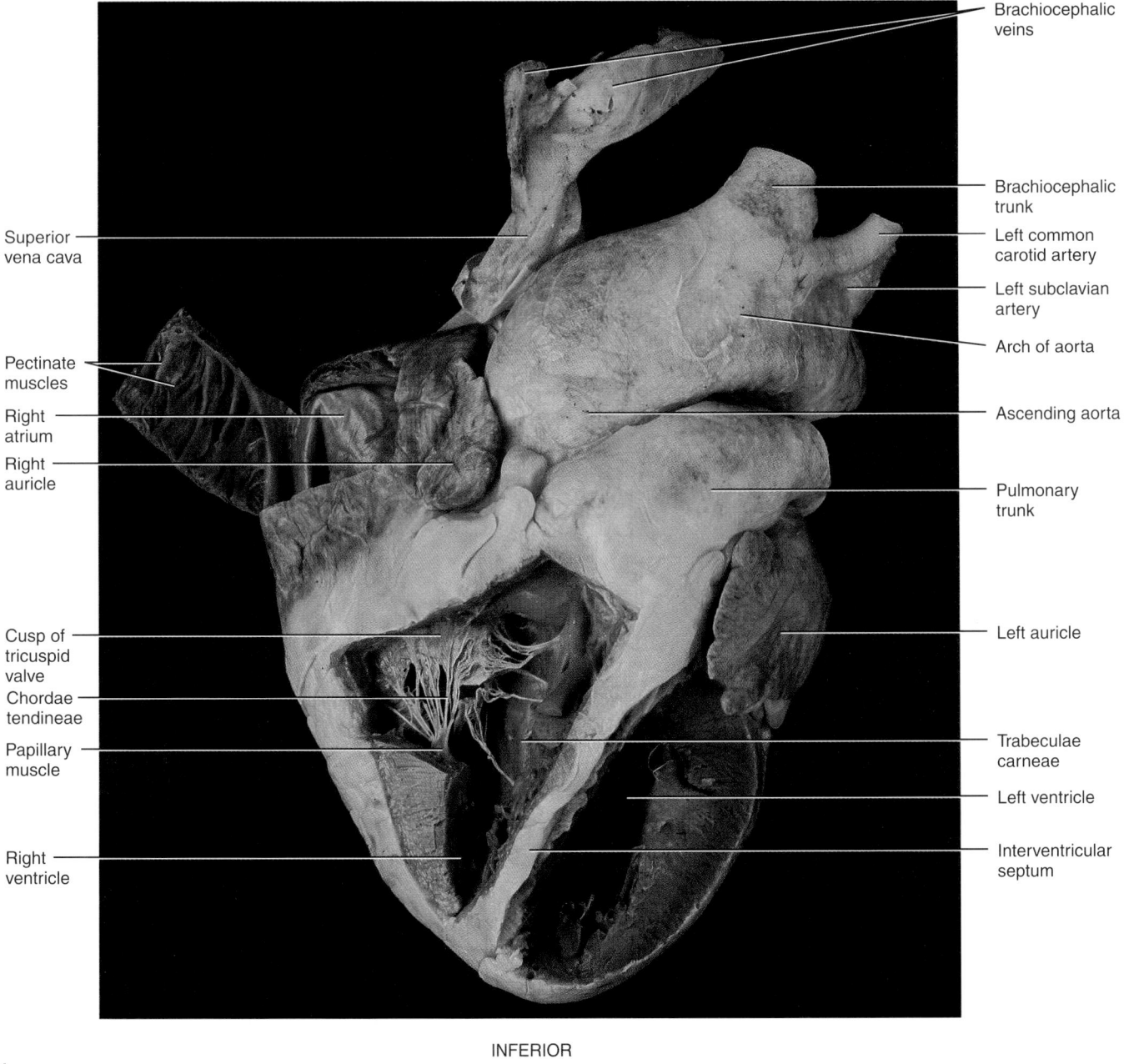

Brachiocephalic
veins

Brachiocephalic
trunk

Left common
carotid artery

Left subclavian
artery

Arch of aorta

Ascending aorta

Pulmonary
trunk

Left auricle

Trabeculae
carneae

Left ventricle

Interventricular
septum

Superior
vena cava

Pectinate
muscles

Right
atrium

Right
auricle

Cusp of
tricuspid
valve

Chordae
tendineae

Papillary
muscle

Right
ventricle

INFERIOR

FIGURE 6.4 | *Heart, partially sectioned*    (a) Anterior view of human heart

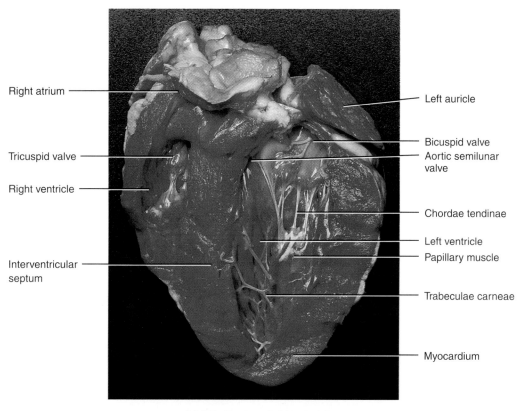

Right atrium

Tricuspid valve

Right ventricle

Interventricular
septum

Left auricle

Bicuspid valve

Aortic semilunar
valve

Chordae tendinae

Left ventricle

Papillary muscle

Trabeculae carneae

Myocardium

(b) Anterior view of sheep heart

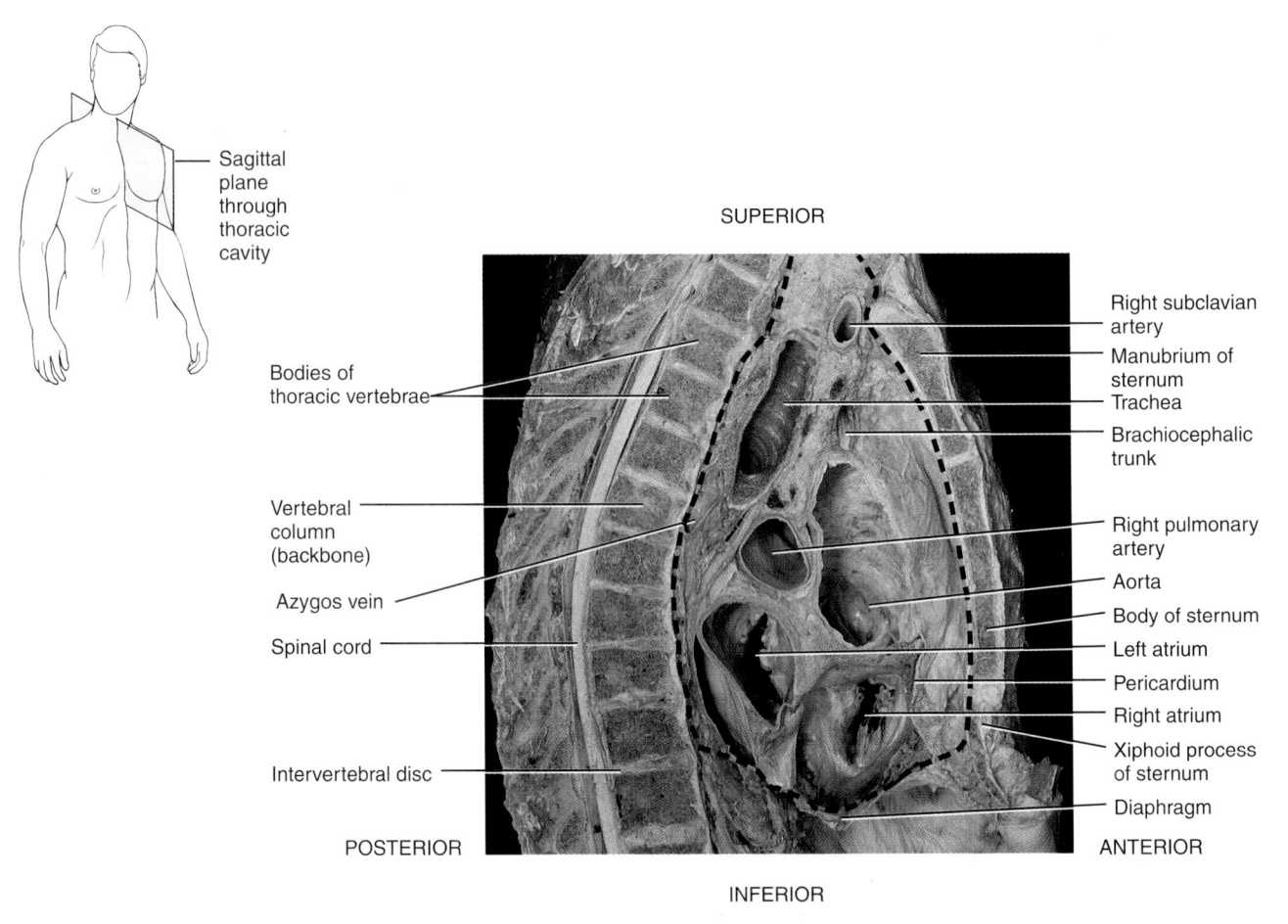

Sagittal plane through thoracic cavity

SUPERIOR

Bodies of thoracic vertebrae

Vertebral column (backbone)

Azygos vein

Spinal cord

Intervertebral disc

POSTERIOR

Right subclavian artery

Manubrium of sternum

Trachea

Brachiocephalic trunk

Right pulmonary artery

Aorta

Body of sternum

Left atrium

Pericardium

Right atrium

Xiphoid process of sternum

Diaphragm

ANTERIOR

INFERIOR

Sagittal section

FIGURE 6.5 | *Heart*

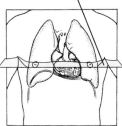

Transverse plane
through thoracic cavity

ANTERIOR

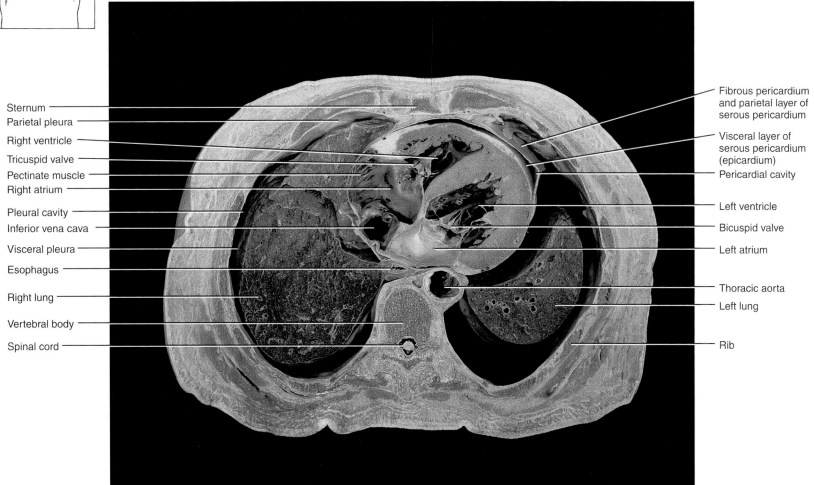

Sternum

Parietal pleura

Right ventricle

Tricuspid valve

Pectinate muscle

Right atrium

Pleural cavity

Inferior vena cava

Visceral pleura

Esophagus

Right lung

Vertebral body

Spinal cord

Fibrous pericardium
and parietal layer of
serous pericardium

Visceral layer of
serous pericardium
(epicardium)

Pericardial cavity

Left ventricle

Bicuspid valve

Left atrium

Thoracic aorta

Left lung

Rib

POSTERIOR

Inferior view of transverse section

FIGURE 6.6 | *Heart*

91

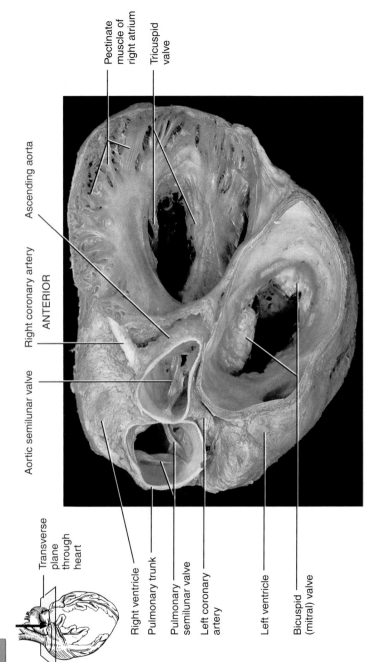

Pectinate muscle of right atrium

Tricuspid valve

Ascending aorta

Right coronary artery
ANTERIOR

Aortic semilunar valve

Right ventricle

Pulmonary trunk

Pulmonary semilunar valve

Left coronary artery

Left ventricle

Bicuspid (mitral) valve

POSTERIOR

Superior view

Transverse plane through heart

F I G U R E   6 . 7   |   *Heart valves*

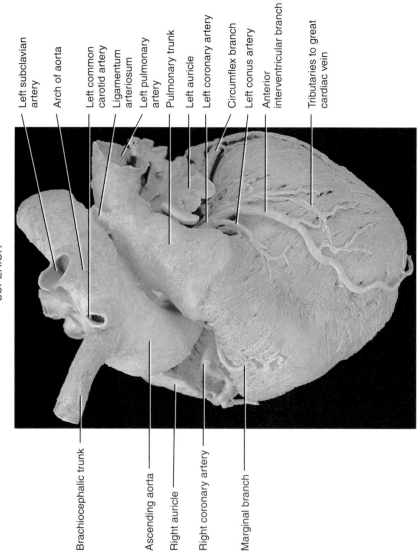

Left subclavian artery

Arch of aorta

Left common carotid artery

Ligamentum arteriosum

Left pulmonary artery

Pulmonary trunk

Left auricle

Left coronary artery

Circumflex branch

Left conus artery

Anterior interventricular branch

Tributaries to great cardiac vein

SUPERIOR

Brachiocephalic trunk

Ascending aorta

Right auricle

Right coronary artery

Marginal branch

INFERIOR

Anterior view

F I G U R E   6 . 8   |   *Blood supply to the heart*

SUPERIOR

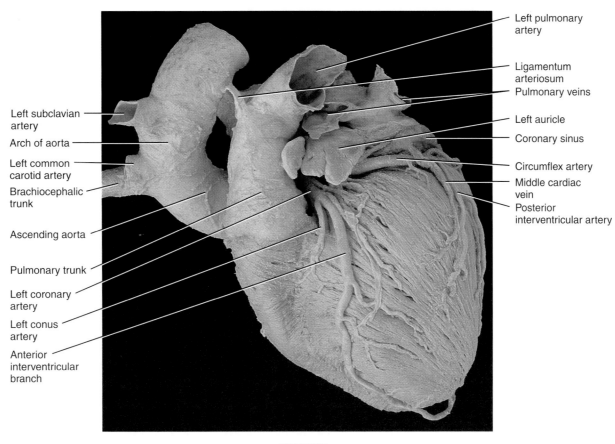

Left pulmonary artery

Ligamentum arteriosum

Pulmonary veins

Left subclavian artery

Left auricle

Arch of aorta

Coronary sinus

Left common carotid artery

Circumflex artery

Brachiocephalic trunk

Middle cardiac vein

Posterior interventricular artery

Ascending aorta

Pulmonary trunk

Left coronary artery

Left conus artery

Anterior interventricular branch

INFERIOR

Anterolateral view

FIGURE 6.9 | *Blood supply to the heart*

SUPERIOR

Larynx

Left common carotid artery

Left vertebral artery

Left subclavian artery

Arch of aorta

Left primary bronchus

Thoracic aorta

Diaphragm

Left adrenal (suprarenal) gland

Splenic artery

Left renal artery

Left kidney

Abdominal aorta

Inferior mesenteric artery

Psoas major muscle

Sigmoid colon

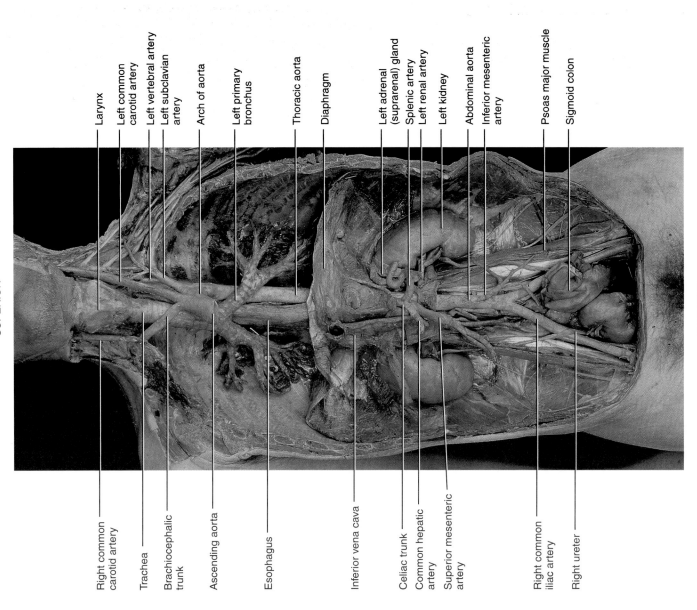

Right common carotid artery

Trachea

Brachiocephalic trunk

Ascending aorta

Esophagus

Inferior vena cava

Celiac trunk

Common hepatic artery

Superior mesenteric artery

Right common iliac artery

Right ureter

INFERIOR

Anterior view

FIGURE 6.10 | *Arteries of the thorax, abdomen, and pelvis*

SUPERIOR

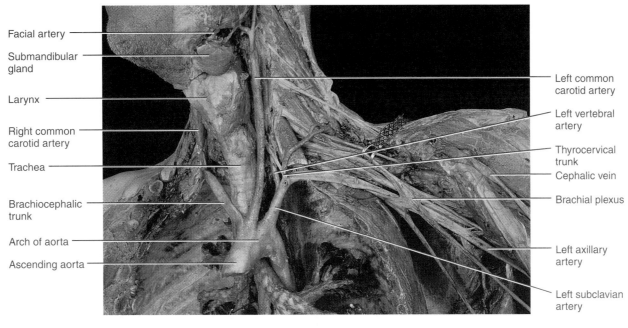

Facial artery

Submandibular
gland

Larynx

Right common
carotid artery

Trachea

Brachiocephalic
trunk

Arch of aorta

Ascending aorta

Left common
carotid artery

Left vertebral
artery

Thyrocervical
trunk

Cephalic vein

Brachial plexus

Left axillary
artery

Left subclavian
artery

INFERIOR

Anterior view

FIGURE  6.11 | *Arteries of the neck
and shoulders*

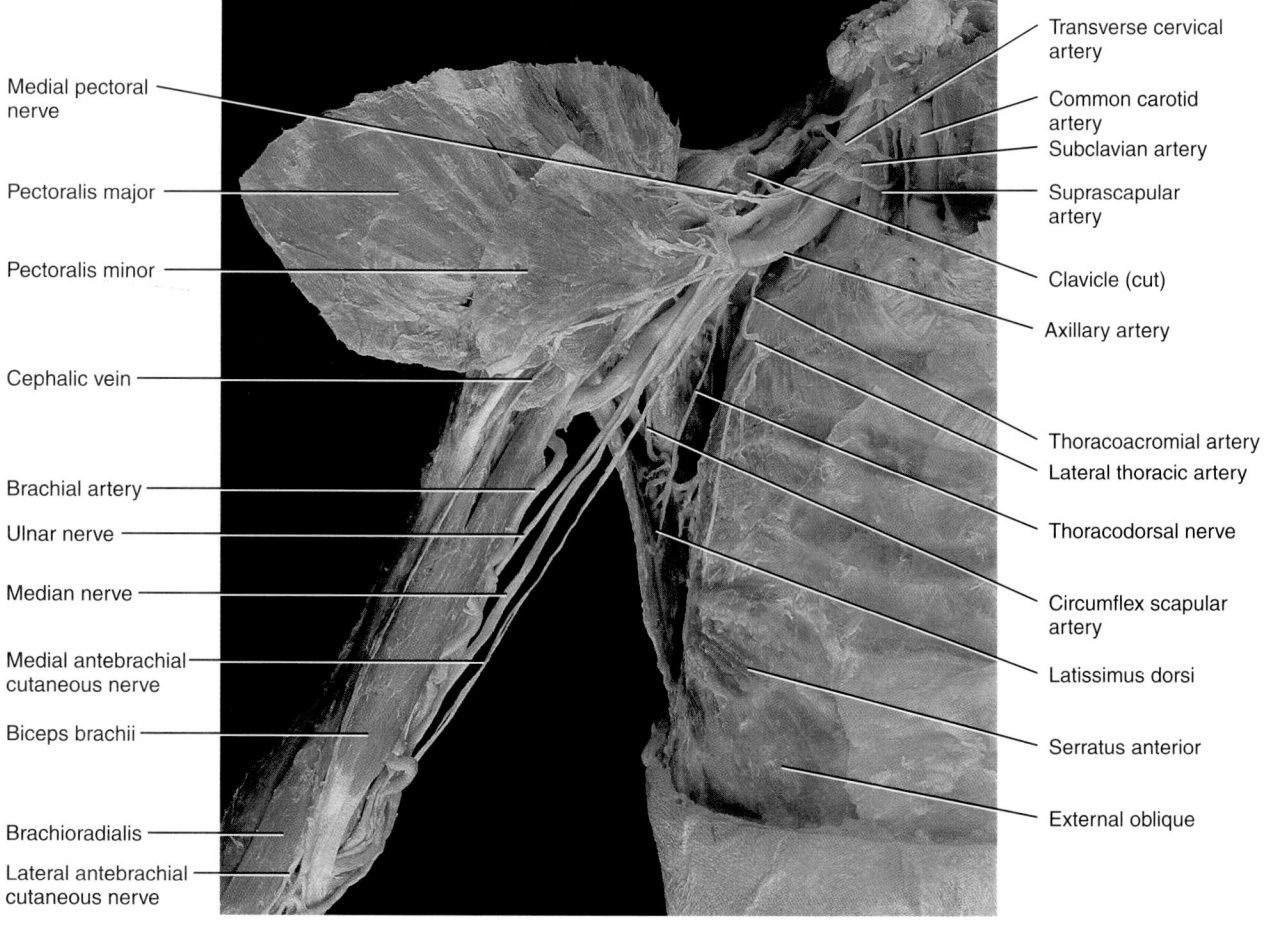

Transverse cervical artery

Common carotid artery

Subclavian artery

Suprascapular artery

Clavicle (cut)

Axillary artery

Thoracoacromial artery

Lateral thoracic artery

Thoracodorsal nerve

Circumflex scapular artery

Latissimus dorsi

Serratus anterior

External oblique

Medial pectoral nerve

Pectoralis major

Pectoralis minor

Cephalic vein

Brachial artery

Ulnar nerve

Median nerve

Medial antebrachial cutaneous nerve

Biceps brachii

Brachioradialis

Lateral antebrachial cutaneous nerve

INFERIOR

Anterior view

FIGURE 6.12 | *Arteries of the neck and right upper limb*

SUPERIOR

Brachial artery

Anterior interosseous artery

Ulnar nerve

Median nerve

Radial artery

Ulnar artery

Hypothenar muscles

Thenar muscles

Superficial palmar arch

Common palmar digital artery

Proper palmar digital artery

LATERAL

MEDIAL

INFERIOR

Anterior view

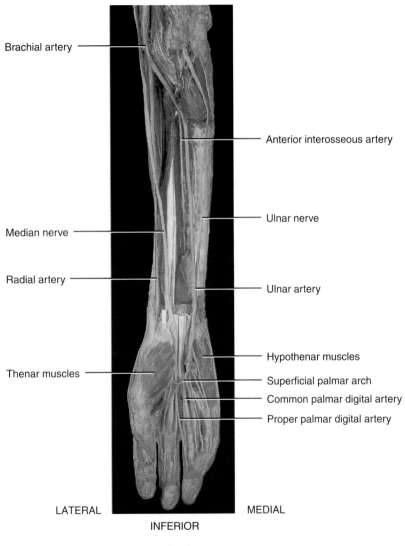

FIGURE 6.13 | *Arteries of the right forearm, wrist, and hand*

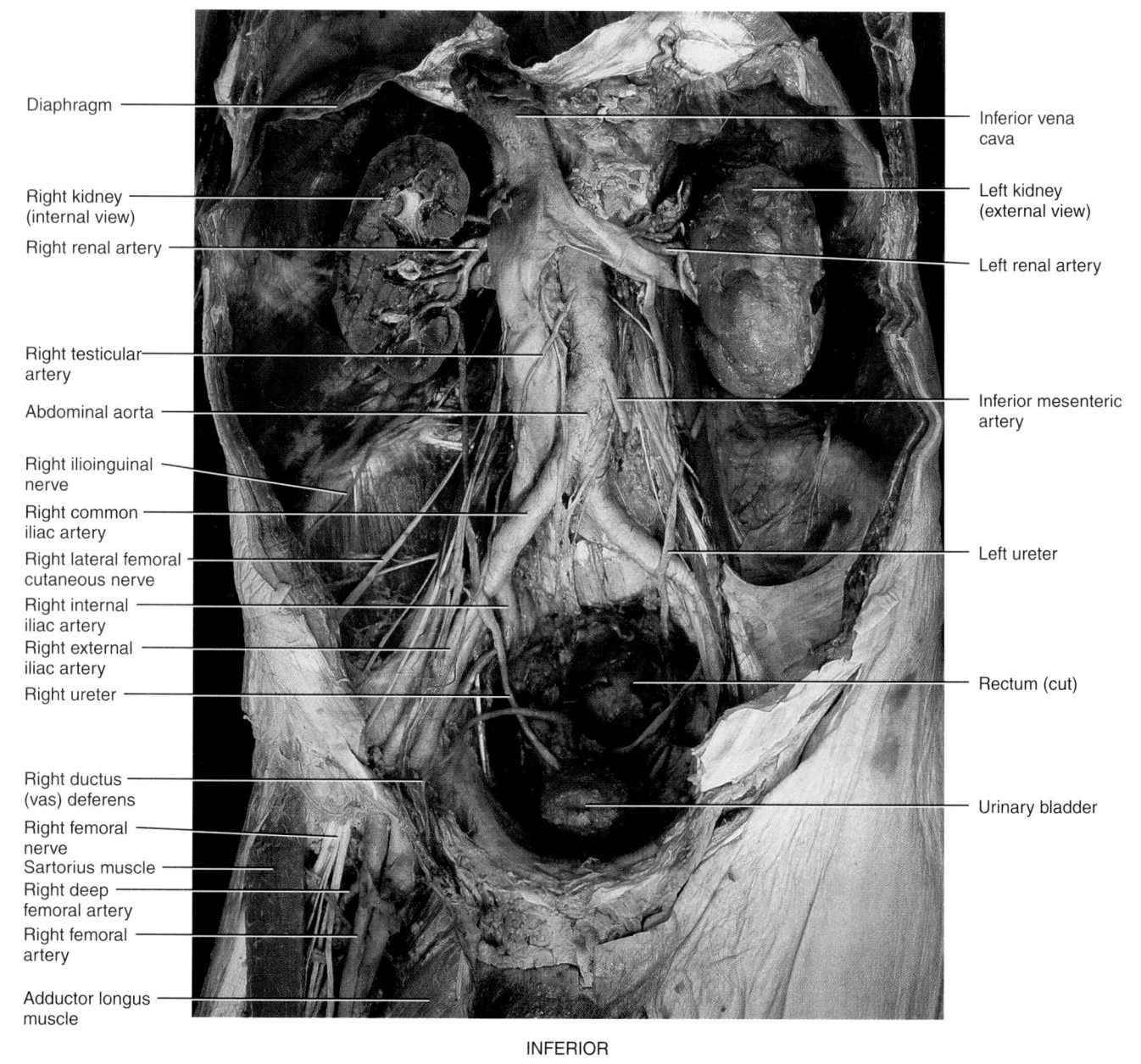

Diaphragm

Right kidney
(internal view)

Right renal artery

Right testicular
artery

Abdominal aorta

Right ilioinguinal
nerve

Right common
iliac artery

Right lateral femoral
cutaneous nerve

Right internal
iliac artery

Right external
iliac artery

Right ureter

Right ductus
(vas) deferens

Right femoral
nerve

Sartorius muscle

Right deep
femoral artery

Right femoral
artery

Adductor longus
muscle

Inferior vena
cava

Left kidney
(external view)

Left renal artery

Inferior mesenteric
artery

Left ureter

Rectum (cut)

Urinary bladder

INFERIOR

Anterior view

FIGURE   6 . 1 4   |   *Arteries of the abdomen*
*and pelvis*

SUPERIOR

Esophagus in
esophageal hiatus

Inferior phrenic
artery

Diaphragm

Inferior vena cava
(cut)

Left adrenal
(suprarenal) gland

Celiac trunk

Splenic artery

Common hepatic
artery

Left renal artery

Superior
mesenteric artery

Right renal vein
(cut)

Right ureter

Middle colic artery

Inferior
mesenteric artery

Right colic artery

Left colic artery

Abdominal aorta

Ileocolic artery

Sigmoid artery

Superior rectal
artery

Right common
iliac artery

Lateral femoral
cutaneous nerve

Sigmoid colon

Right external
iliac artery

Right external
iliac vein

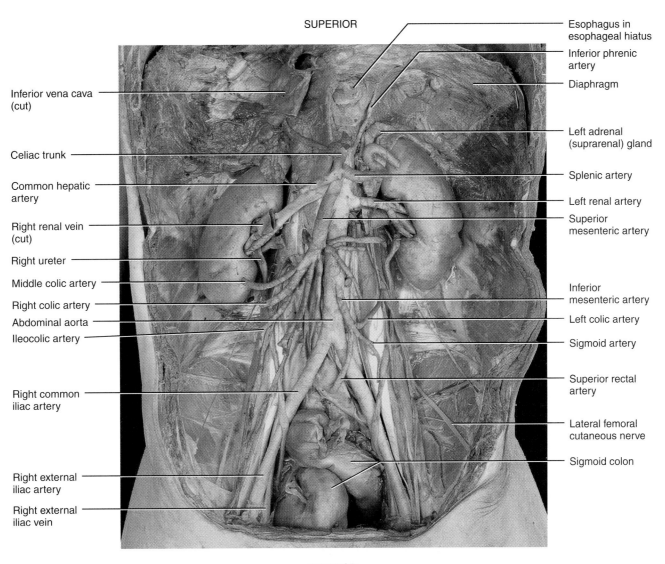

INFERIOR

Anterior view

FIGURE  6 . 1 5  | *Arteries of the abdomen*
*and pelvis*

99

SUPERIOR

Liver

Transverse
colon

Hepatic
portal vein

Descending
colon

Superior
mesenteric vein

Splenic vein
(cut)

Right colic vein

Ileocolic vein

Superior
mesenteric
artery

Ascending colon

Jejunum

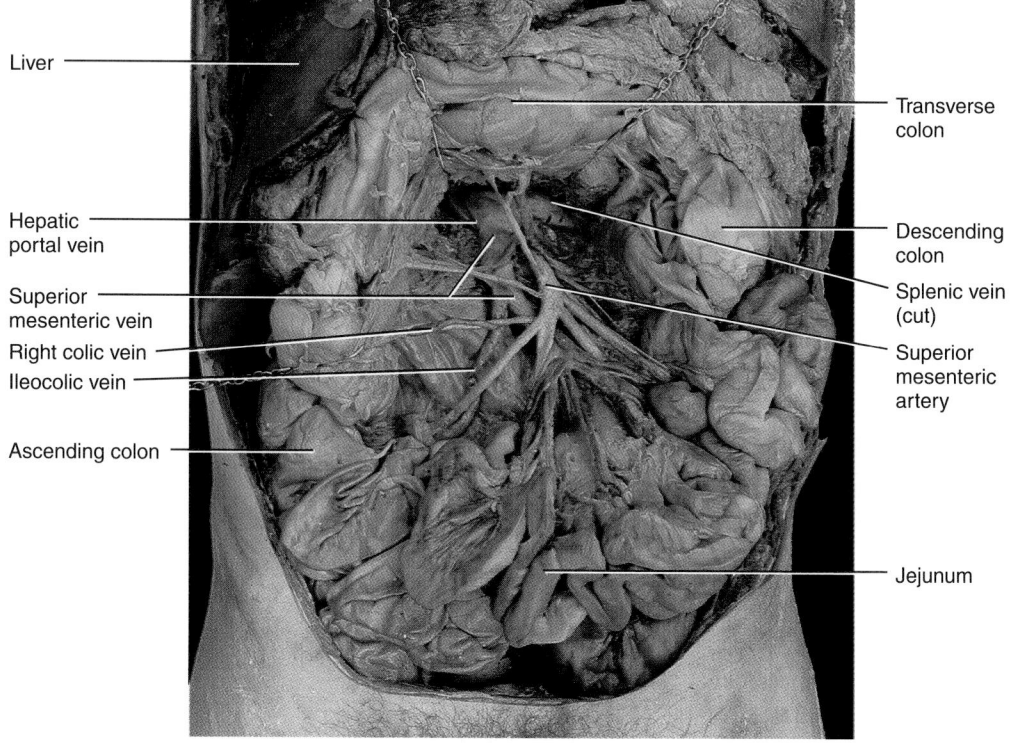

INFERIOR

Anterior view

FIGURE  6.16  |  *Tributaries of the hepatic portal vein and*
*branches of the superior mesenteric artery*

SUPERIOR

External iliac vein

External iliac artery

Inguinal ligament

Tensor fasciae latae muscle

Femoral nerve

Deep femoral artery

Lateral circumflex vein

Femoral vein

Femoral artery

Muscular branch

Iliotibial tract

Vastus lateralis muscle

Sartorius muscle

Rectus femoris muscle

LATERAL

Right ureter

Urinary bladder

Ductus (vas) deferens

Spermatic cord

Pectineus muscle

Inguinal lymph node

Penis

Adductor longus muscle

Gracilis muscle

MEDIAL

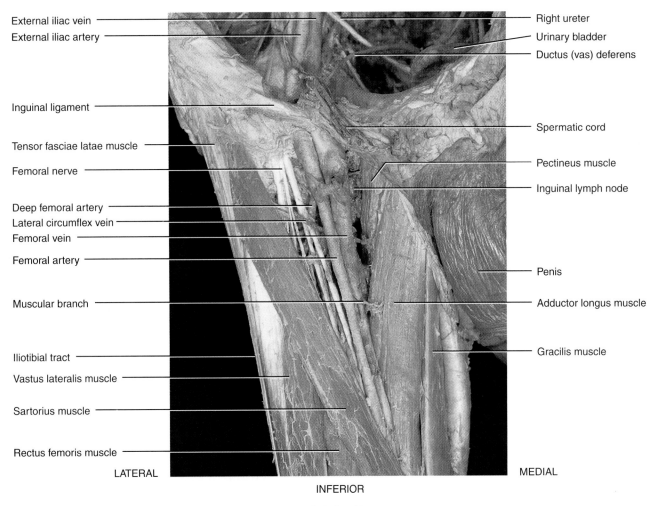

INFERIOR

Anterior view

FIGURE 6.17 | *Blood vessels of the right thigh*

SUPERIOR

Internal
jugular vein

Left brachiocephalic
vein

Internal
thoracic vein

Heart

Subclavian
vein

Axillary vein

Brachial vein

Basilic vein

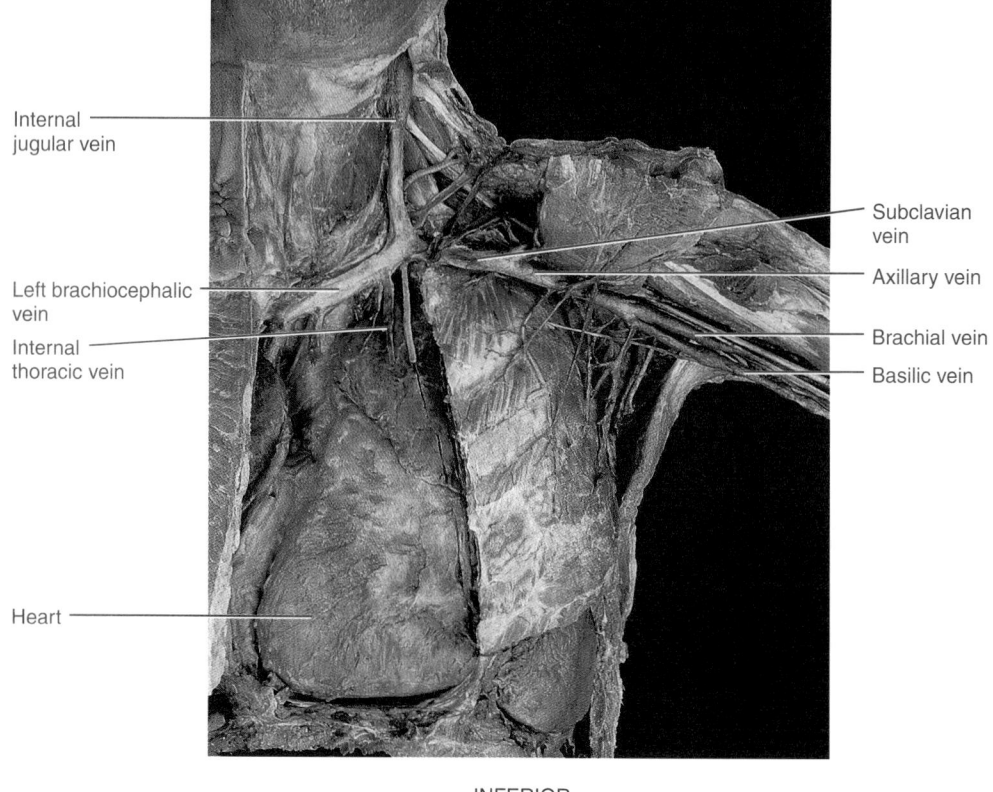

INFERIOR

Anterior view

FIGURE 6.18 | *Veins of the neck
and left shoulder*

SUPERIOR

Basilic vein

Basilic vein

Median antebrachial
vein

Cephalic vein

Biceps brachii muscle

Median cubital vein

Accessory cephalic vein

Cephalic vein

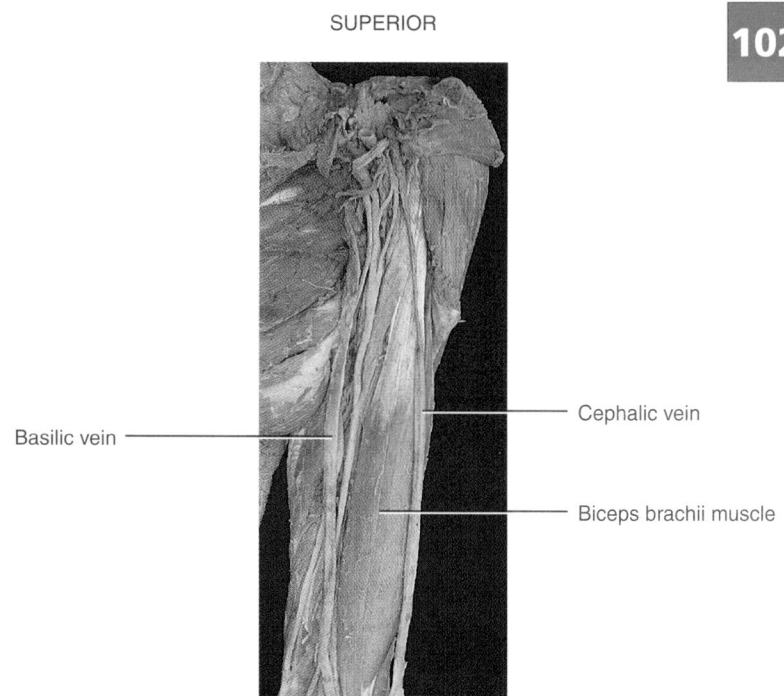

MEDIAL

LATERAL

INFERIOR

Anterior view

FIGURE 6.19 | *Veins of the left arm
and forearm*

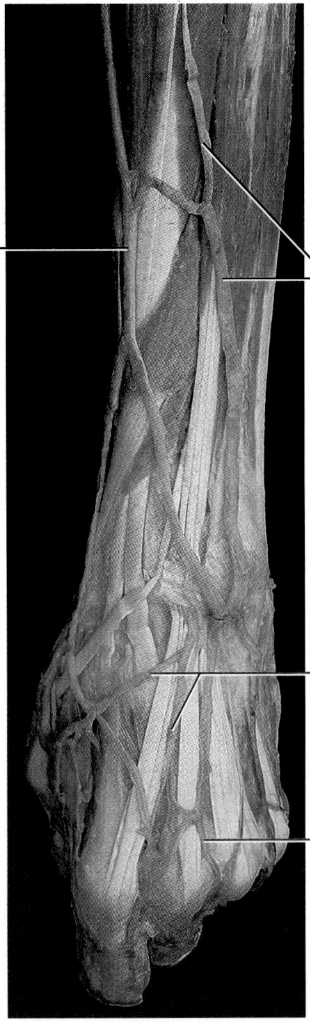

SUPERIOR

Cephalic vein

Accessory
cephalic vein

Dorsal venous
arch

Dorsal digital
vein

MEDIAL

LATERAL

INFERIOR

Posterior view

FIGURE 6.20 | *Veins of the left forearm and hand*

103

FIGURE 6.21 | *Veins of the abdomen and pelvis*

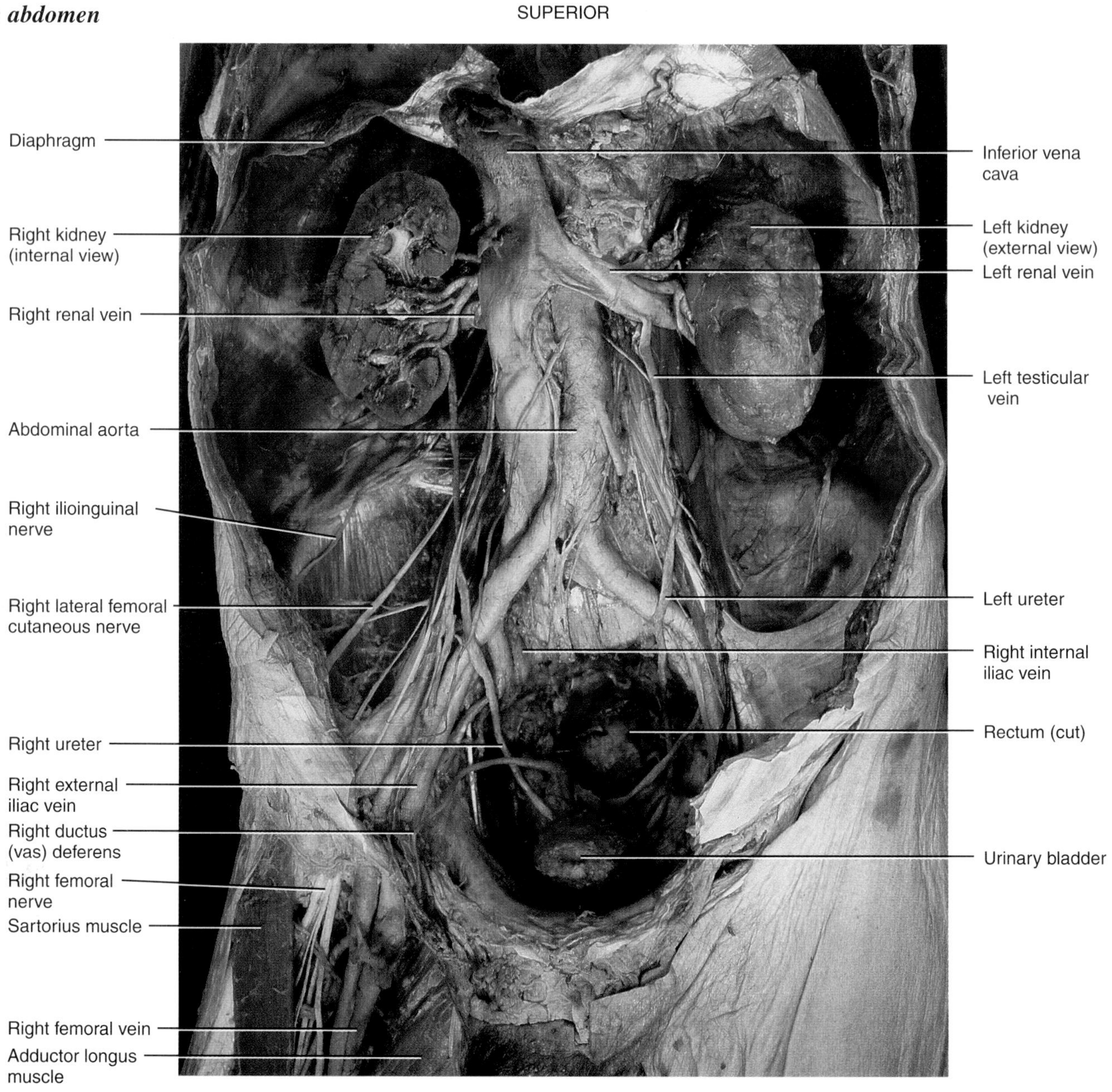

SUPERIOR

Diaphragm

Right kidney
(internal view)

Right renal vein

Abdominal aorta

Right ilioinguinal
nerve

Right lateral femoral
cutaneous nerve

Right ureter

Right external
iliac vein

Right ductus
(vas) deferens

Right femoral
nerve

Sartorius muscle

Right femoral vein

Adductor longus
muscle

Inferior vena
cava

Left kidney
(external view)

Left renal vein

Left testicular
vein

Left ureter

Right internal
iliac vein

Rectum (cut)

Urinary bladder

INFERIOR

Anterior view

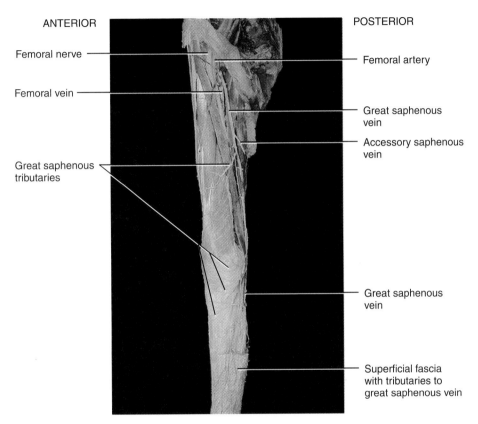

ANTERIOR                          POSTERIOR

Femoral nerve

Femoral vein

Great saphenous
tributaries

Femoral artery

Great saphenous
vein

Accessory saphenous
vein

Great saphenous
vein

Superficial fascia
with tributaries to
great saphenous vein

(a) Medial view of veins of the thigh and leg

FIGURE 6.22 | *Veins of the right lower limb*

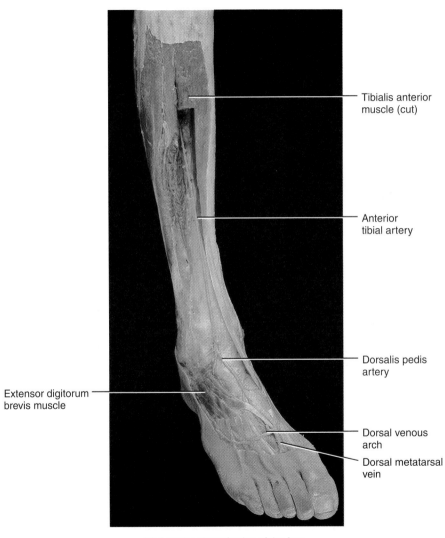

Tibialis anterior
muscle (cut)

Anterior
tibial artery

Dorsalis pedis
artery

Extensor digitorum
brevis muscle

Dorsal venous
arch

Dorsal metatarsal
vein

(b) Anterior view of veins of the foot

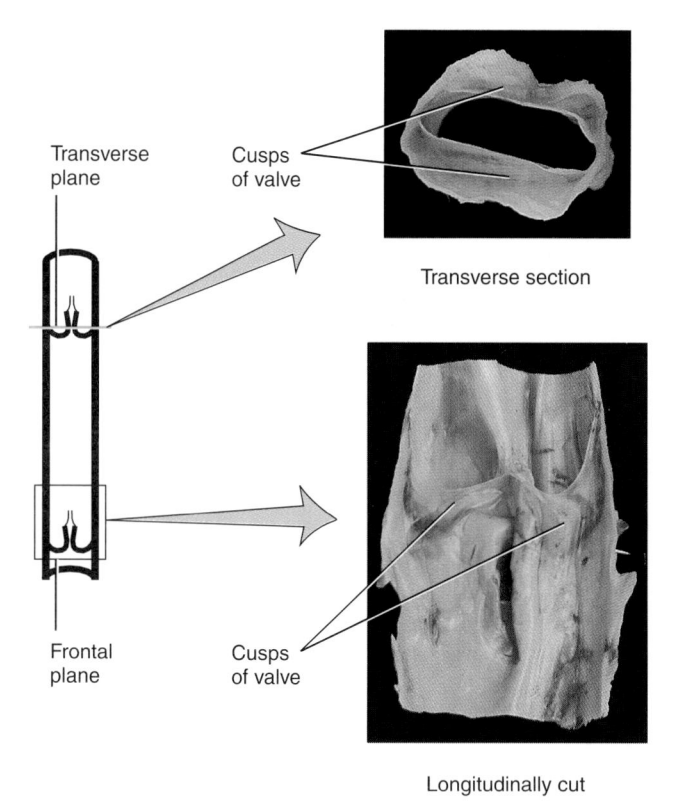

Transverse
plane

Cusps
of valve

Transverse section

Frontal
plane

Cusps
of valve

Longitudinally cut

FIGURE  6.23  |  *Valves in a vein*

# UNIT SEVEN | *The Lymphatic System*

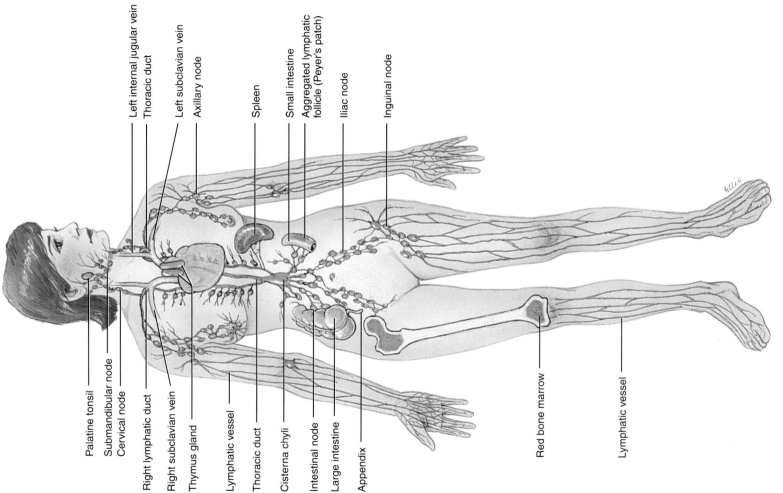

Palatine tonsil

Submandibular node

Cervical node

Right lymphatic duct

Right subclavian vein

Thymus gland

Lymphatic vessel

Thoracic duct

Cisterna chyli

Intestinal node

Large intestine

Appendix

Left internal jugular vein

Thoracic duct

Left subclavian vein

Axillary node

Spleen

Small intestine

Aggregated lymphatic follicle (Peyer's patch)

Iliac node

Inguinal node

Red bone marrow

Lymphatic vessel

Anterior view

FIGURE 7.1 | *Lymphatic system*

FIGURE 7.2 | *Thymus gland*

SUPERIOR

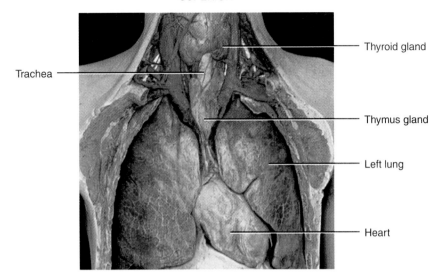

Thyroid gland

Trachea

Thymus gland

Left lung

Heart

INFERIOR

(a) Anterior view of thymus gland

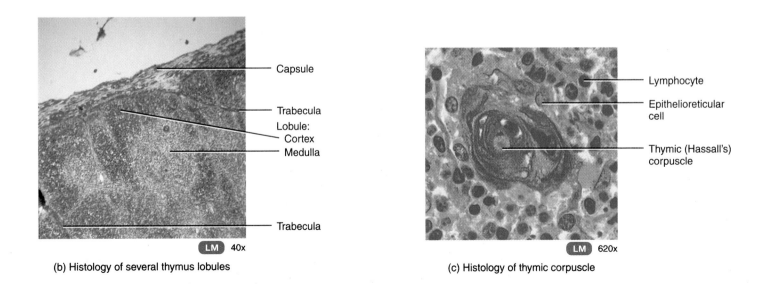

Capsule

Trabecula

Lobule:
Cortex
Medulla

Trabecula

LM 40x

(b) Histology of several thymus lobules

Lymphocyte

Epithelioreticular
cell

Thymic (Hassall's)
corpuscle

LM 620x

(c) Histology of thymic corpuscle

FIGURE 7.3 | *Spleen*   FIGURE 7.4 | *Lymph node*

SUPERIOR

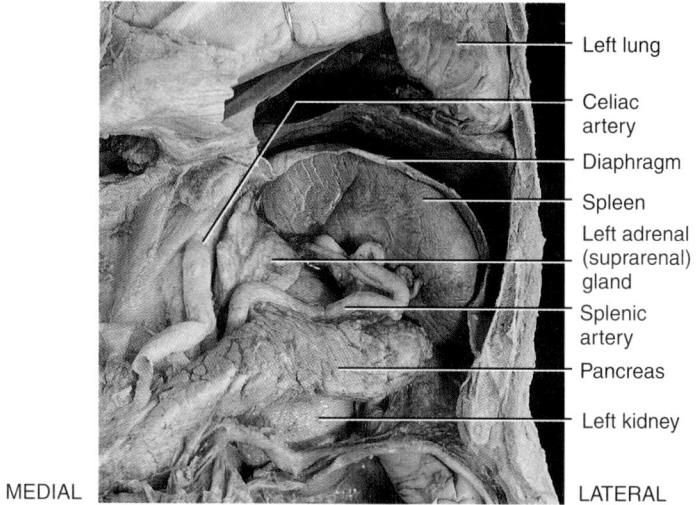

Left lung

Celiac artery

Diaphragm

Spleen

Left adrenal (suprarenal) gland

Splenic artery

Pancreas

Left kidney

MEDIAL                                    LATERAL

INFERIOR

(a) Anterior view of the spleen

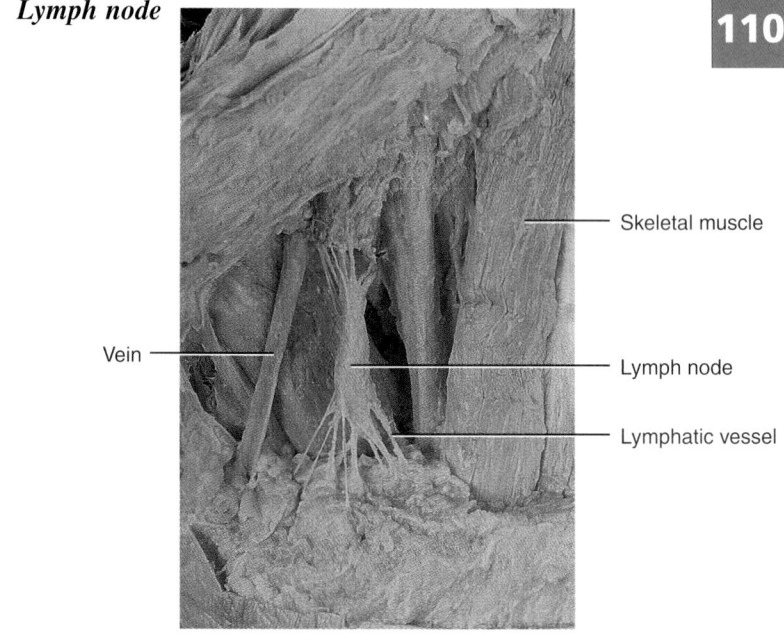

Skeletal muscle

Vein

Lymph node

Lymphatic vessel

(a) Anterior view of a lymph node

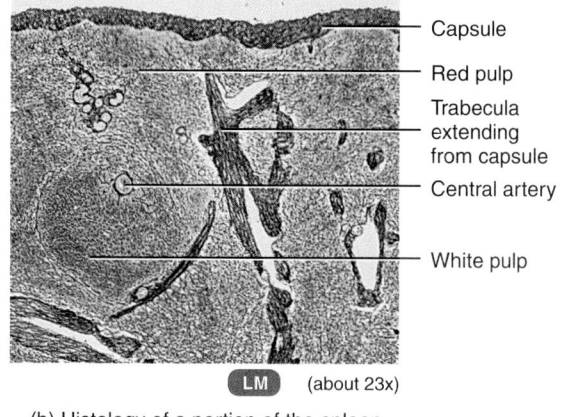

Capsule

Red pulp

Trabecula extending from capsule

Central artery

White pulp

**LM** (about 23x)

(b) Histology of a portion of the spleen

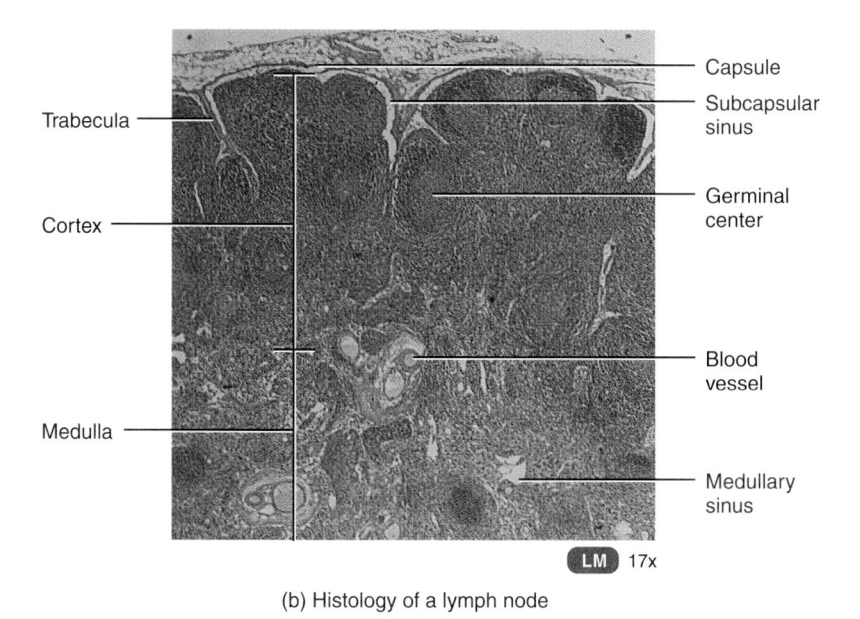

Trabecula

Cortex

Medulla

Capsule

Subcapsular sinus

Germinal center

Blood vessel

Medullary sinus

**LM** 17x

(b) Histology of a lymph node

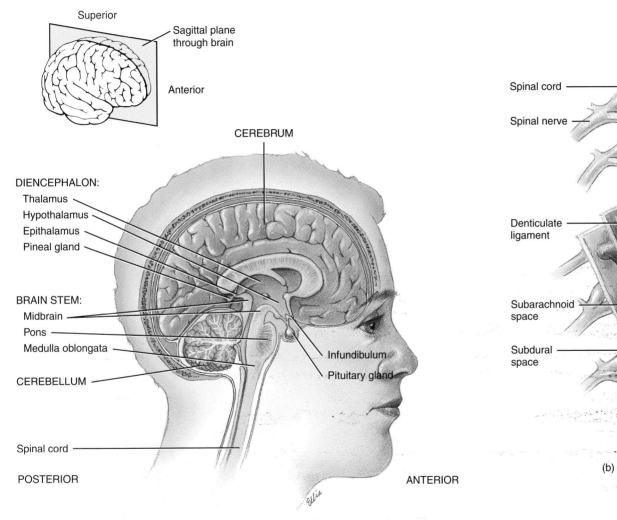

Superior

Sagittal plane
through brain

Anterior

CEREBRUM

DIENCEPHALON:
Thalamus
Hypothalamus
Epithalamus
Pineal gland

BRAIN STEM:
Midbrain
Pons
Medulla oblongata

CEREBELLUM

Infundibulum

Pituitary gland

Spinal cord

POSTERIOR

ANTERIOR

(a) Sagittal section of brain

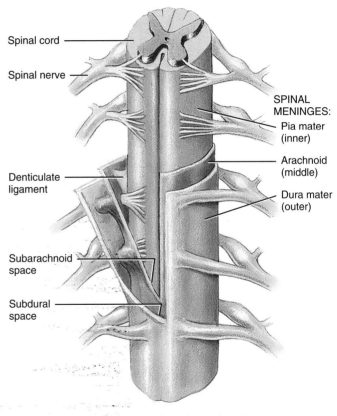

Spinal cord

Spinal nerve

SPINAL
MENINGES:
Pia mater
(inner)

Arachnoid
(middle)

Denticulate
ligament

Dura mater
(outer)

Subarachnoid
space

Subdural
space

(b) Sections through spinal cord

FIGURE 8.1 | *Brain and*
*spinal cord*

SUPERIOR

Fourth ventricle

Glossopharyngeal (IX)
and vagus (X) nerves

Accessory (XI) nerve

Fasciculus gracilis

Fasciculus cuneatus

Dura mater
and arachnoid

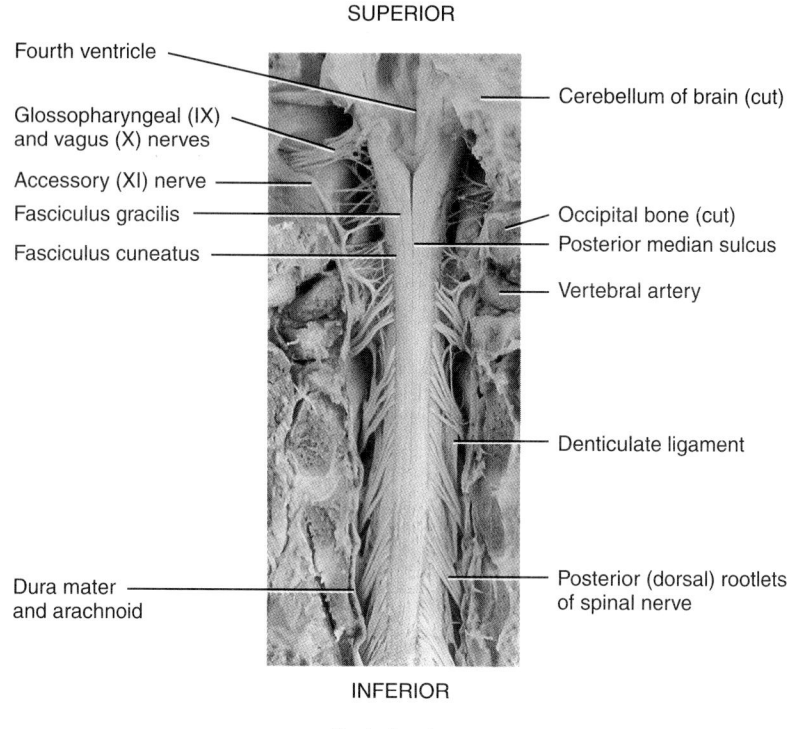

Cerebellum of brain (cut)

Occipital bone (cut)
Posterior median sulcus

Vertebral artery

Denticulate ligament

Posterior (dorsal) rootlets
of spinal nerve

INFERIOR

Posterior view

FIGURE 8.2 | *Spinal cord, cervical region*

SUPERIOR

Conus medullaris

Dura mater
and arachnoid

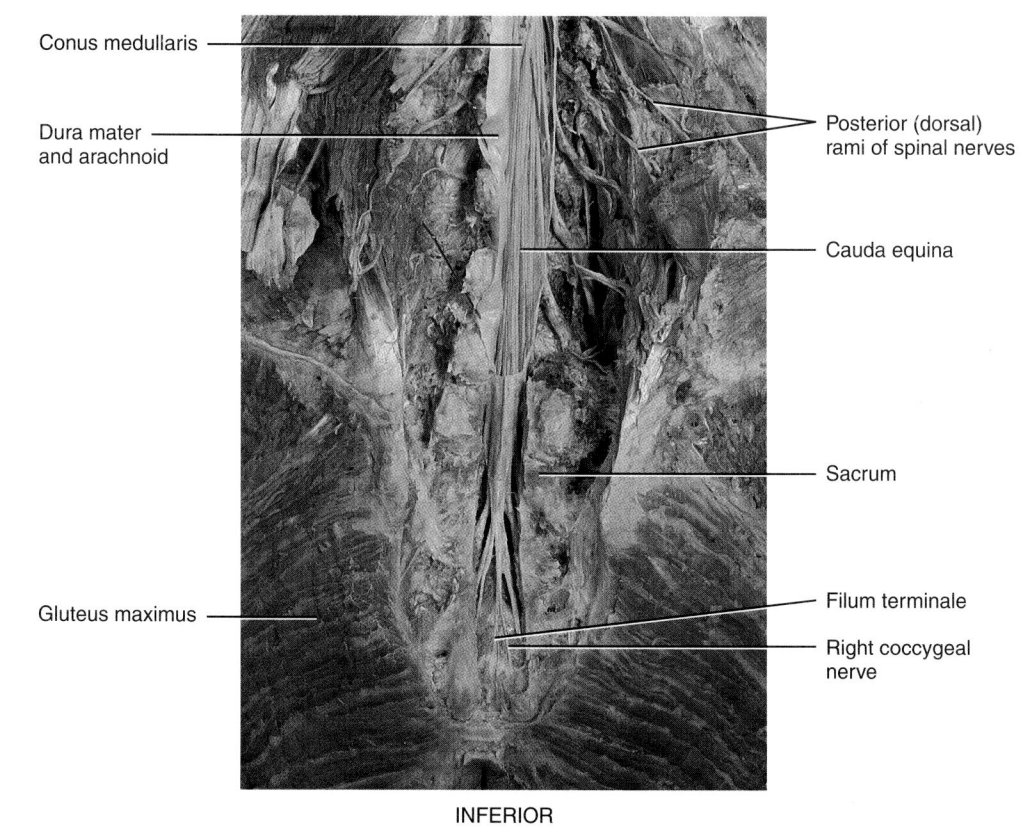

Posterior (dorsal)
rami of spinal nerves

Cauda equina

Sacrum

Gluteus maximus

Filum terminale

Right coccygeal
nerve

INFERIOR

Posterior view

FIGURE 8.3 | *Spinal cord, lumbosacral region*

POSTERIOR

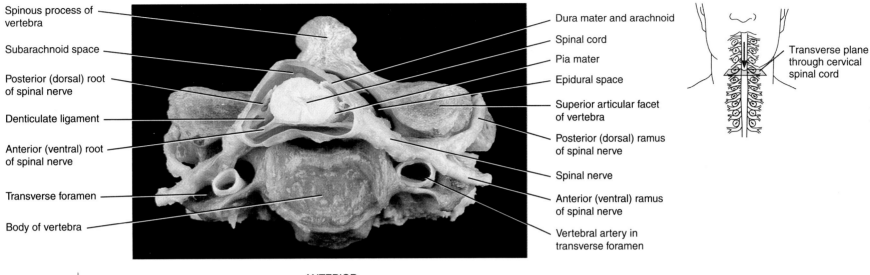

Spinous process of vertebra

Subarachnoid space

Posterior (dorsal) root of spinal nerve

Denticulate ligament

Anterior (ventral) root of spinal nerve

Transverse foramen

Body of vertebra

Dura mater and arachnoid

Spinal cord

Pia mater

Epidural space

Superior articular facet of vertebra

Posterior (dorsal) ramus of spinal nerve

Spinal nerve

Anterior (ventral) ramus of spinal nerve

Vertebral artery in transverse foramen

Transverse plane through cervical spinal cord

ANTERIOR

Transverse section

FIGURE 8.4 | *Spinal cord, cervical region*

POSTERIOR

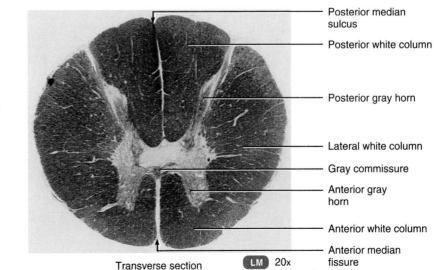

Transverse plane through thoracic spinal cord

Posterior median sulcus

Posterior white column

Posterior gray horn

Lateral white column

Gray commissure

Anterior gray horn

Anterior white column

Anterior median fissure

Transverse section     LM  20x

ANTERIOR

FIGURE 8.5 | *Spinal cord, thoracic region*

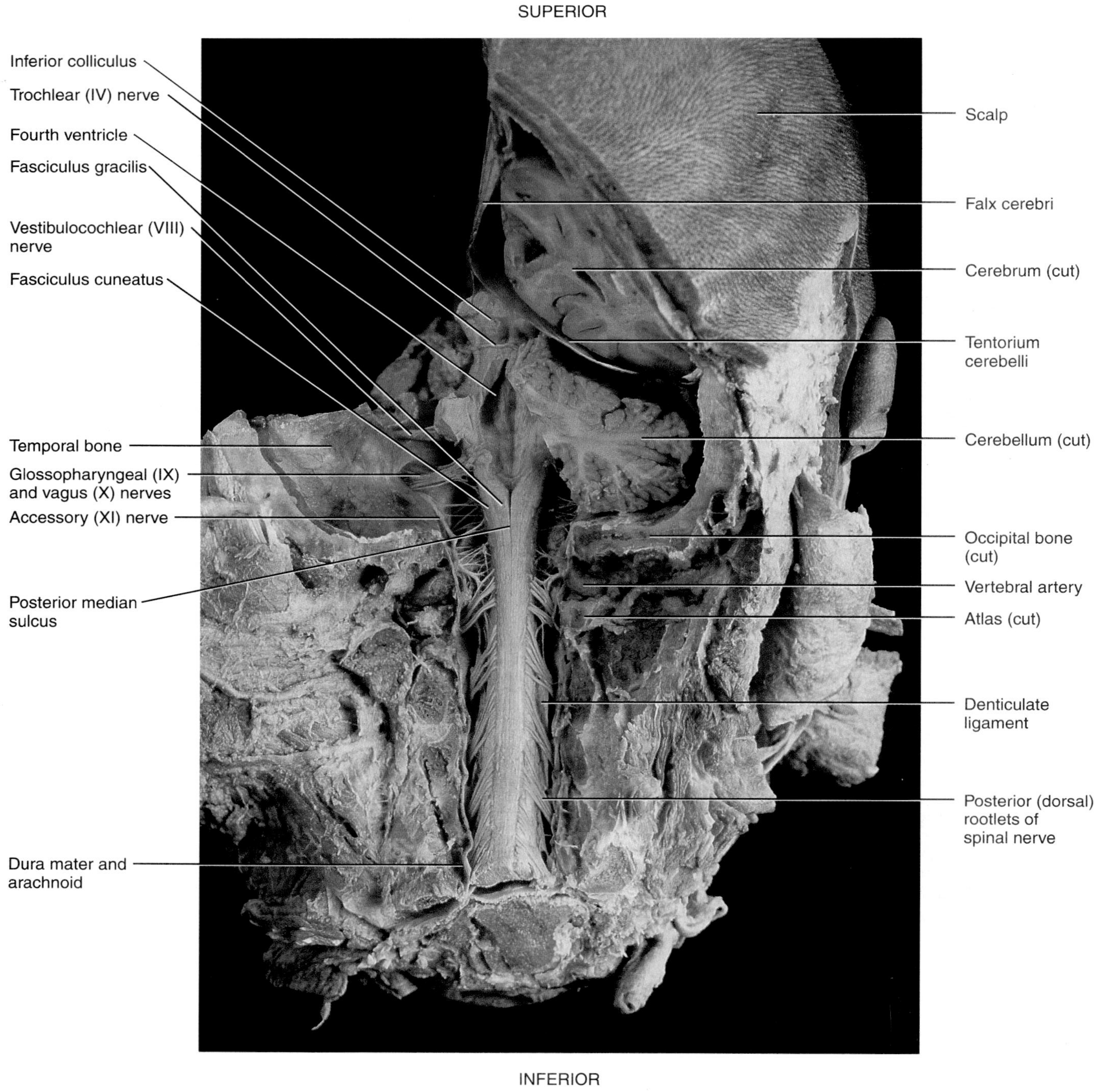

Labels (top, left to right):
Scalp
Falx cerebri
Cerebrum (cut)
Tentorium cerebelli
Cerebellum (cut)
Occipital bone (cut)
Vertebral artery
Atlas (cut)
Denticulate ligament
Posterior (dorsal) rootlets of spinal nerve

Labels (bottom, left to right):
Inferior colliculus
Trochlear (IV) nerve
Fourth ventricle
Fasciculus gracilis
Vestibulocochlear (VIII) nerve
Fasciculus cuneatus
Temporal bone
Glossopharyngeal (IX) and vagus (X) nerves
Accessory (XI) nerve
Posterior median sulcus
Dura mater and arachnoid

SUPERIOR

INFERIOR

FIGURE 8.6 | *Spinal cord (posterior view) and brain (oblique section)*

114

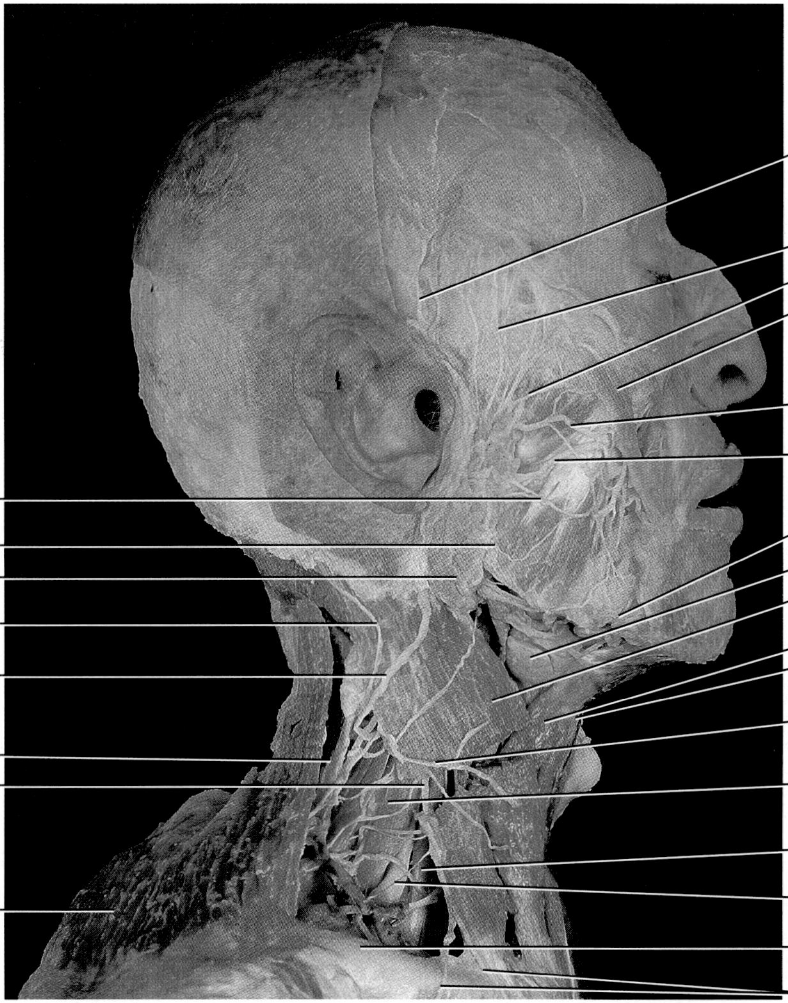

SUPERIOR

POSTERIOR

ANTERIOR

Auriculotemporal
nerve

Facial (VII) nerve:

Temporal branch

Zygomatic branch

Zygomaticus major
muscle

Transverse
facial artery

Duct of paratoid
gland

Facial artery

Submandibular gland

Sternocleidomastoid
muscle (cut)

Omohyoid muscle

Sternohyoid muscle

Transverse cervical
nerve

Middle scalene
muscle

Anterior scalene
muscle

Superior trunk of
brachial plexus

Clavicle

Supraclavicular
nerves

Facial (VII) Nerve:

Buccal branch

Mandibular branch

Parotid gland

Lesser occipital
nerve

Great auricular
nerve

Accessory (XI) nerve

Supraclavicular
nerve

Trapezius muscle

INFERIOR

Right lateral view

FIGURE 8.7 | *Facial nerves and
cervical plexus*

115

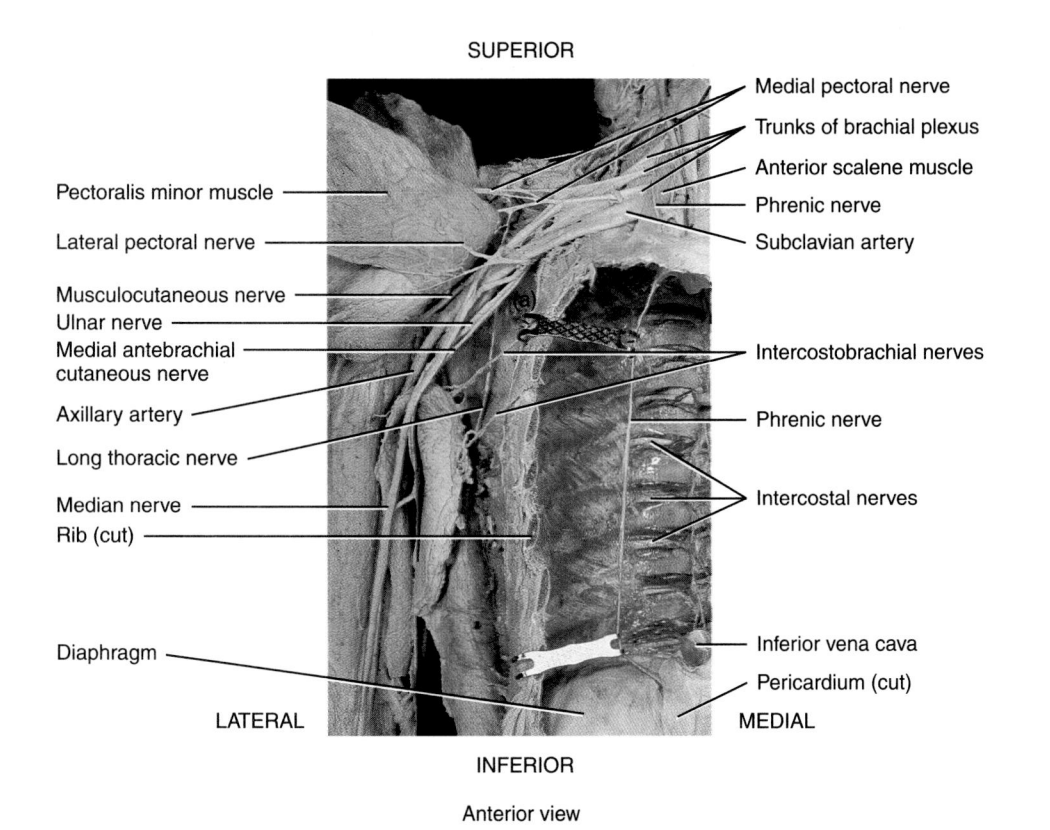

SUPERIOR

Medial pectoral nerve

Trunks of brachial plexus

Anterior scalene muscle

Pectoralis minor muscle

Phrenic nerve

Lateral pectoral nerve

Subclavian artery

Musculocutaneous nerve

Ulnar nerve

Medial antebrachial
cutaneous nerve

Intercostobrachial nerves

Axillary artery

Phrenic nerve

Long thoracic nerve

Median nerve

Intercostal nerves

Rib (cut)

Diaphragm

Inferior vena cava

Pericardium (cut)

LATERAL

MEDIAL

INFERIOR

Anterior view

FIGURE 8.8 | *Brachial plexus*

SUPERIOR

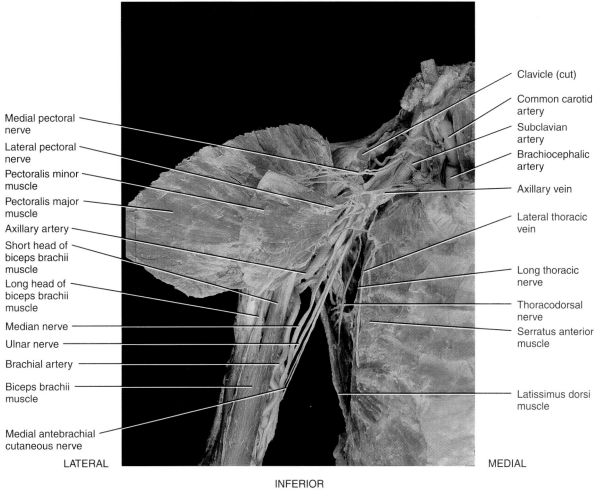

Medial pectoral nerve

Lateral pectoral nerve

Pectoralis minor muscle

Pectoralis major muscle

Axillary artery

Short head of biceps brachii muscle

Long head of biceps brachii muscle

Median nerve

Ulnar nerve

Brachial artery

Biceps brachii muscle

Medial antebrachial cutaneous nerve

Clavicle (cut)

Common carotid artery

Subclavian artery

Brachiocephalic artery

Axillary vein

Lateral thoracic vein

Long thoracic nerve

Thoracodorsal nerve

Serratus anterior muscle

Latissimus dorsi muscle

LATERAL

MEDIAL

INFERIOR

Anterior view

FIGURE  8.9  | *Brachial plexus*

SUPERIOR

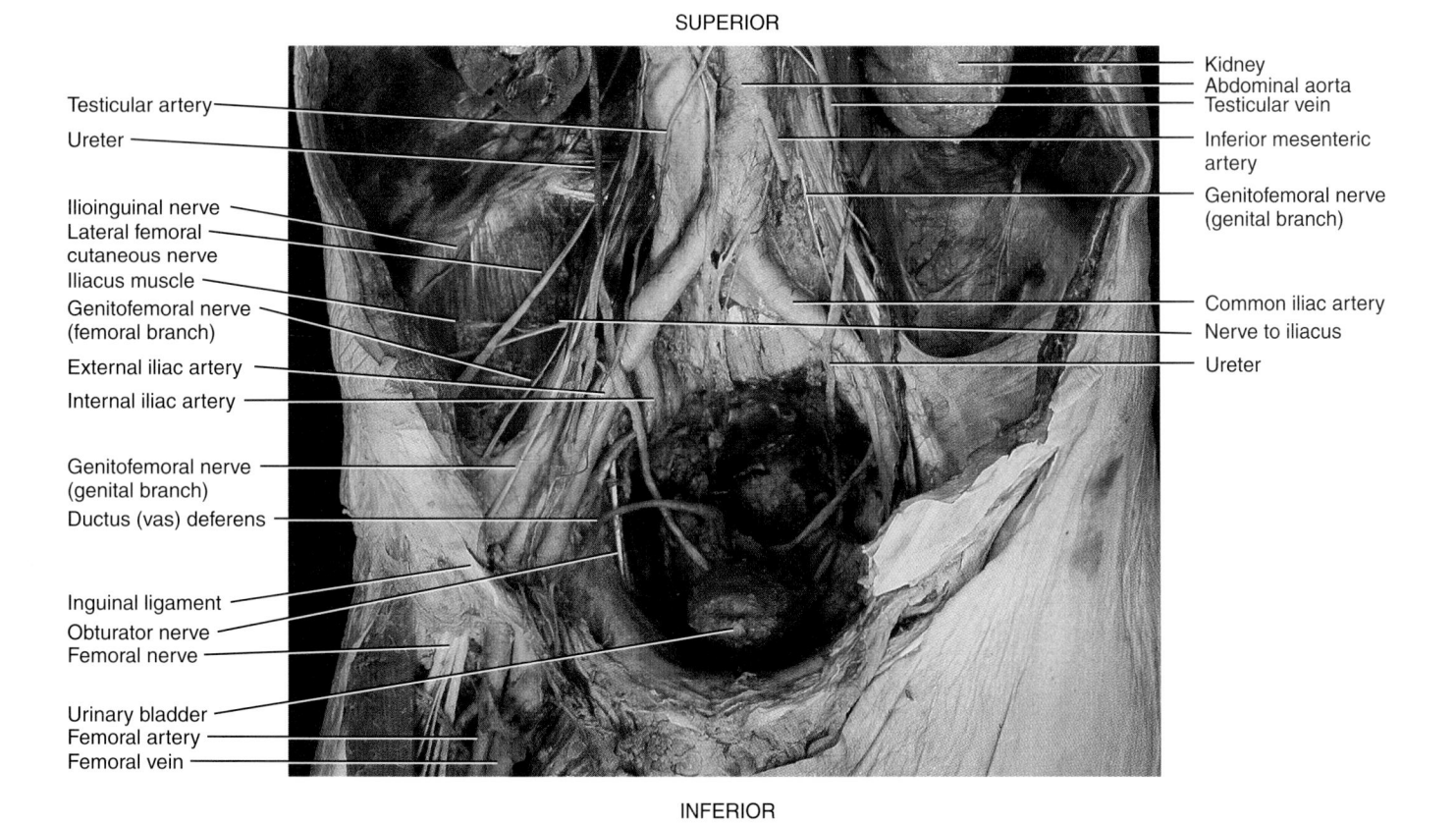

Testicular artery

Ureter

Ilioinguinal nerve
Lateral femoral
cutaneous nerve
Iliacus muscle
Genitofemoral nerve
(femoral branch)

External iliac artery

Internal iliac artery

Genitofemoral nerve
(genital branch)
Ductus (vas) deferens

Inguinal ligament
Obturator nerve
Femoral nerve

Urinary bladder
Femoral artery
Femoral vein

Kidney
Abdominal aorta
Testicular vein
Inferior mesenteric
artery
Genitofemoral nerve
(genital branch)

Common iliac artery
Nerve to iliacus
Ureter

INFERIOR

Anterior view

F I G U R E   8 . 1 0   |   *Lumbar plexus*

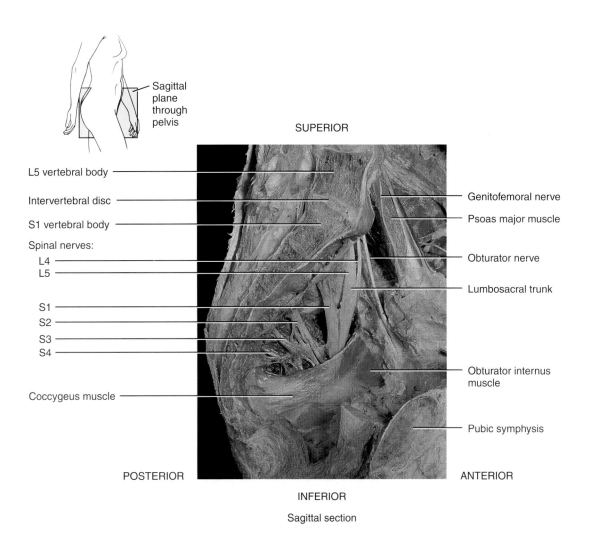

Sagittal plane through pelvis

SUPERIOR

L5 vertebral body

Intervertebral disc

S1 vertebral body

Spinal nerves:
  L4
  L5

  S1
  S2
  S3
  S4

Coccygeus muscle

Genitofemoral nerve

Psoas major muscle

Obturator nerve

Lumbosacral trunk

Obturator internus muscle

Pubic symphysis

POSTERIOR

ANTERIOR

INFERIOR

Sagittal section

FIGURE 8.11 | *Sacral plexus*

SUPERIOR

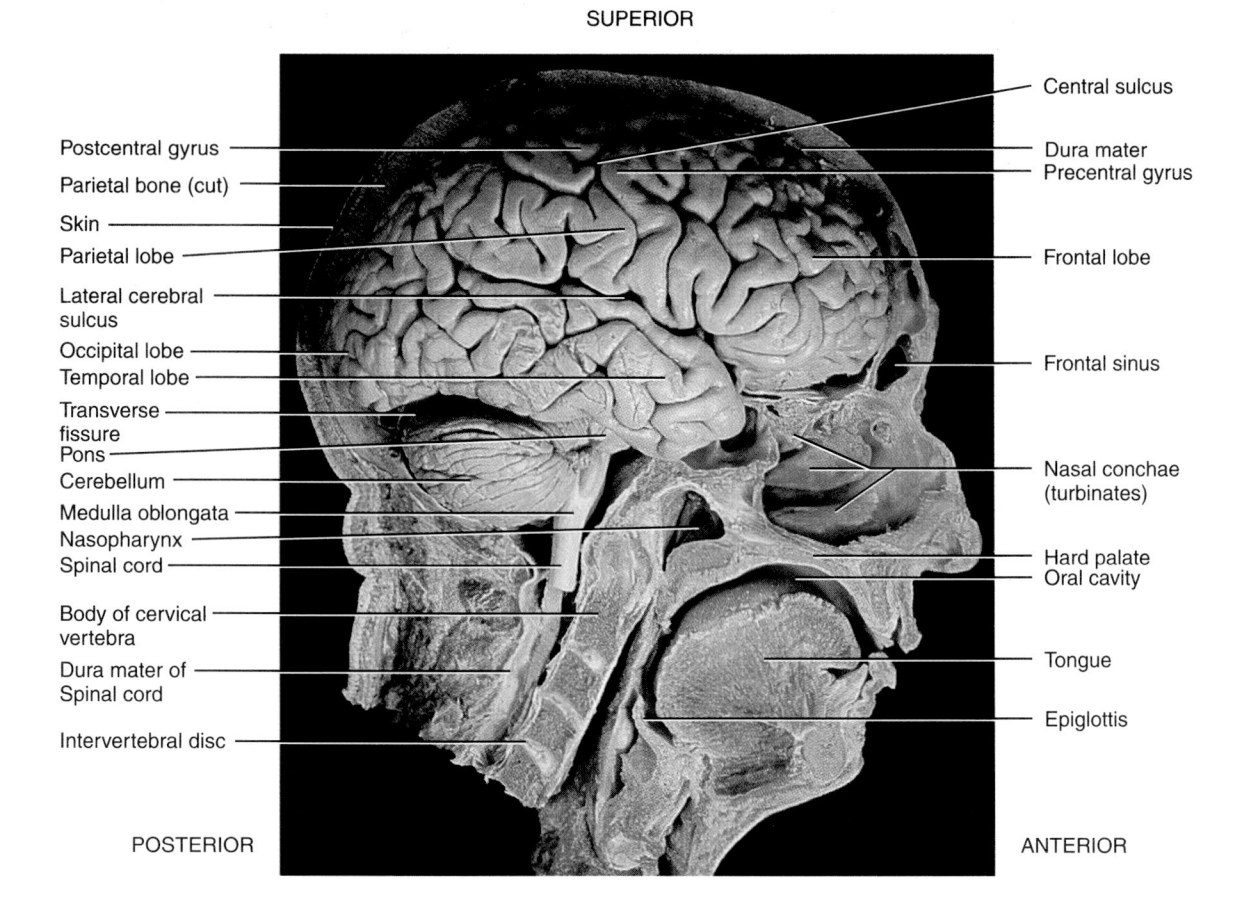

Central sulcus

Postcentral gyrus

Dura mater
Precentral gyrus

Parietal bone (cut)

Skin

Parietal lobe

Frontal lobe

Lateral cerebral
sulcus

Occipital lobe

Temporal lobe

Frontal sinus

Transverse
fissure

Pons

Cerebellum

Nasal conchae
(turbinates)

Medulla oblongata

Nasopharynx

Spinal cord

Hard palate
Oral cavity

Body of cervical
vertebra

Dura mater of
Spinal cord

Tongue

Intervertebral disc

Epiglottis

POSTERIOR

ANTERIOR

INFERIOR

Right lateral view

FIGURE 8.12 | *Brain in cranial cavity*

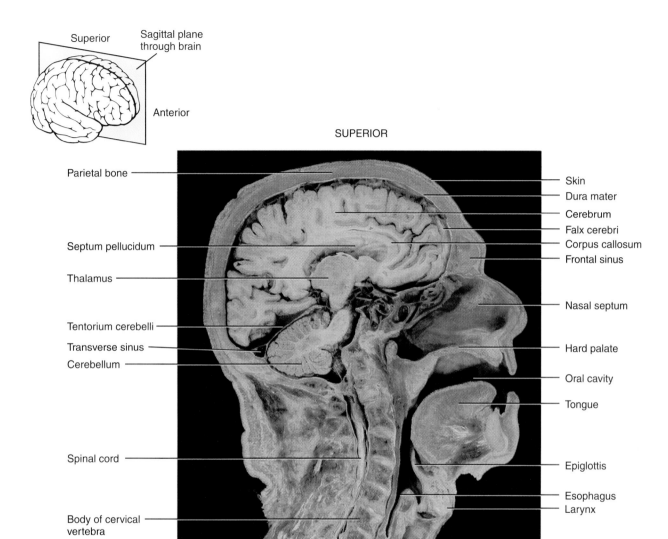

Superior

Sagittal plane
through brain

Anterior

SUPERIOR

Parietal bone

Septum pellucidum

Thalamus

Tentorium cerebelli

Transverse sinus

Cerebellum

Spinal cord

Body of cervical
vertebra

Intervertebral disc

POSTERIOR

Skin

Dura mater

Cerebrum

Falx cerebri

Corpus callosum

Frontal sinus

Nasal septum

Hard palate

Oral cavity

Tongue

Epiglottis

Esophagus

Larynx

Trachea

ANTERIOR

INFERIOR

Sagittal section

FIGURE 8.13 | *Brain and spinal cord*

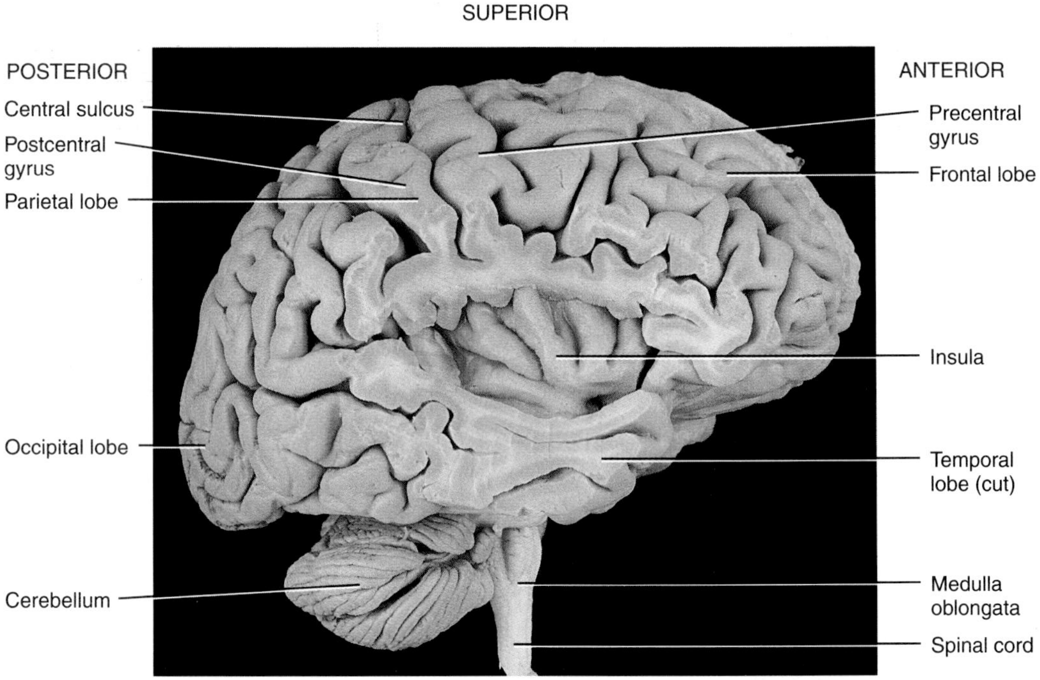

SUPERIOR

POSTERIOR

Central sulcus

Postcentral gyrus

Parietal lobe

Occipital lobe

Cerebellum

ANTERIOR

Precentral gyrus

Frontal lobe

Insula

Temporal lobe (cut)

Medulla oblongata

Spinal cord

INFERIOR

Right lateral view with temporal lobe cut away

FIGURE 8.14 | *Brain*

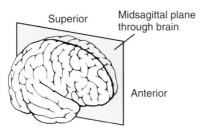

Superior

Midsagittal plane
through brain

Anterior

SUPERIOR

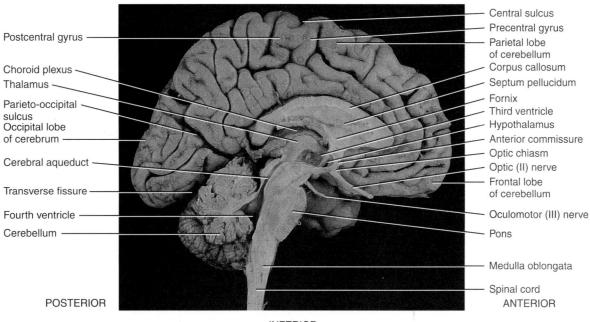

Postcentral gyrus

Choroid plexus

Thalamus

Parieto-occipital
sulcus

Occipital lobe
of cerebrum

Cerebral aqueduct

Transverse fissure

Fourth ventricle

Cerebellum

Central sulcus

Precentral gyrus

Parietal lobe
of cerebellum

Corpus callosum

Septum pellucidum

Fornix

Third ventricle

Hypothalamus

Anterior commissure

Optic chiasm

Optic (II) nerve

Frontal lobe
of cerebellum

Oculomotor (III) nerve

Pons

Medulla oblongata

Spinal cord

POSTERIOR

ANTERIOR

INFERIOR

Midsagittal section

(a) Human brain

FIGURE 8.15 | *Brain*

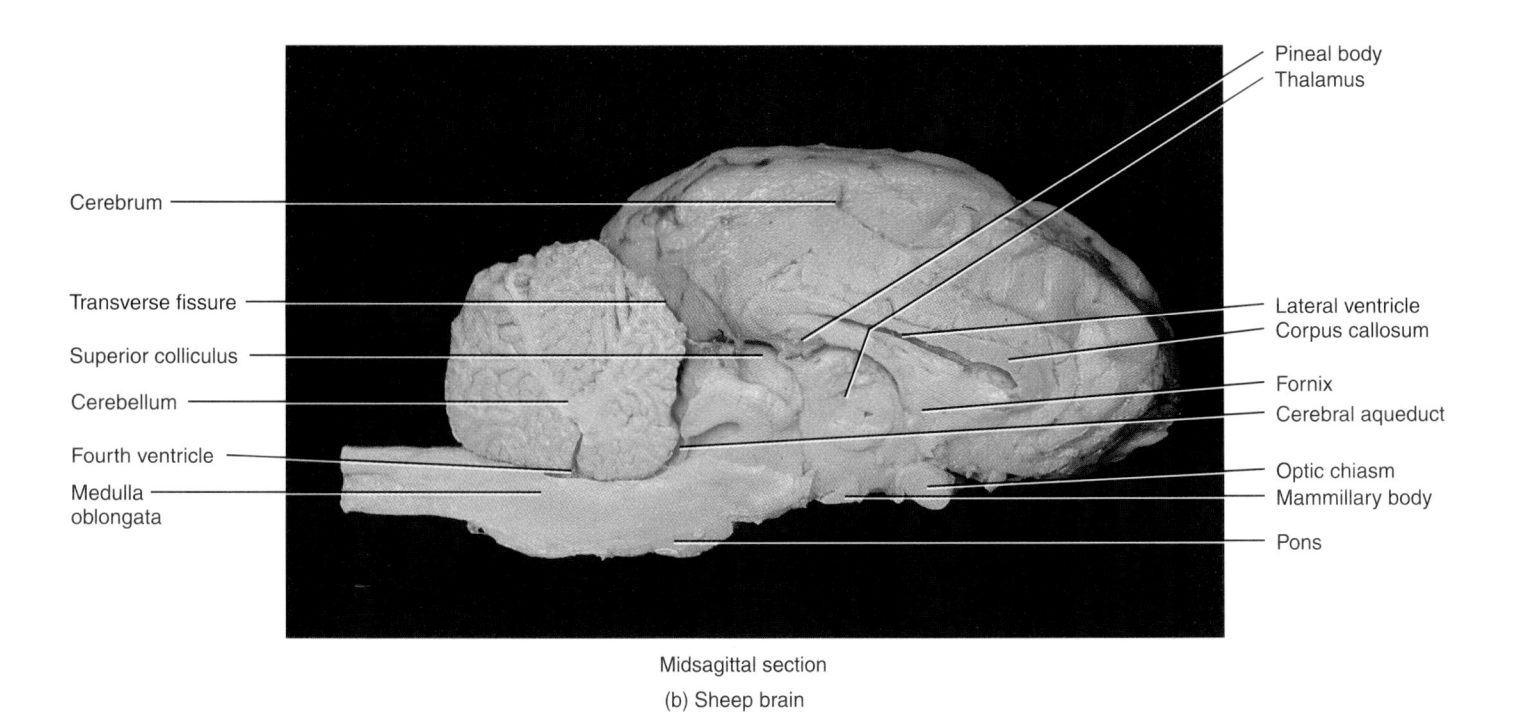

Cerebrum

Transverse fissure

Superior colliculus

Cerebellum

Fourth ventricle

Medulla
oblongata

Pineal body
Thalamus

Lateral ventricle
Corpus callosum

Fornix
Cerebral aqueduct

Optic chiasm
Mammillary body

Pons

Midsagittal section

(b) Sheep brain

FIGURE 8.15 *Brain, continued*

FIGURE 8.16 | *White matter tracts of cerebrum revealed by scooping out gray matter of cerebrum*

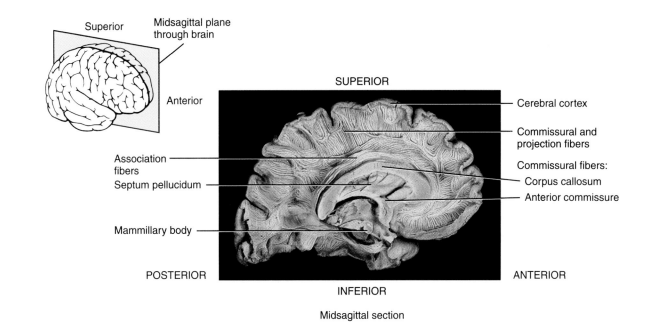

Superior

Midsagittal plane through brain

Anterior

SUPERIOR

Cerebral cortex

Commissural and projection fibers

Commissural fibers:

Corpus callosum

Anterior commissure

Association fibers

Septum pellucidum

Mammillary body

POSTERIOR

ANTERIOR

INFERIOR

Midsagittal section

Superior

Midsagittal plane through brain

Anterior

SUPERIOR

Body of caudate nucleus

Frontal lobe of cerebrum

Head of caudate nucleus

Amygdala

Internal capsule of cerebrum

Tail of caudate nucleus

Occipital lobe of cerebrum

POSTERIOR

ANTERIOR

INFERIOR

Midsagittal section

FIGURE 8.17 | *Basal ganglia revealed by scooping out gray and white matter of cerebrum*

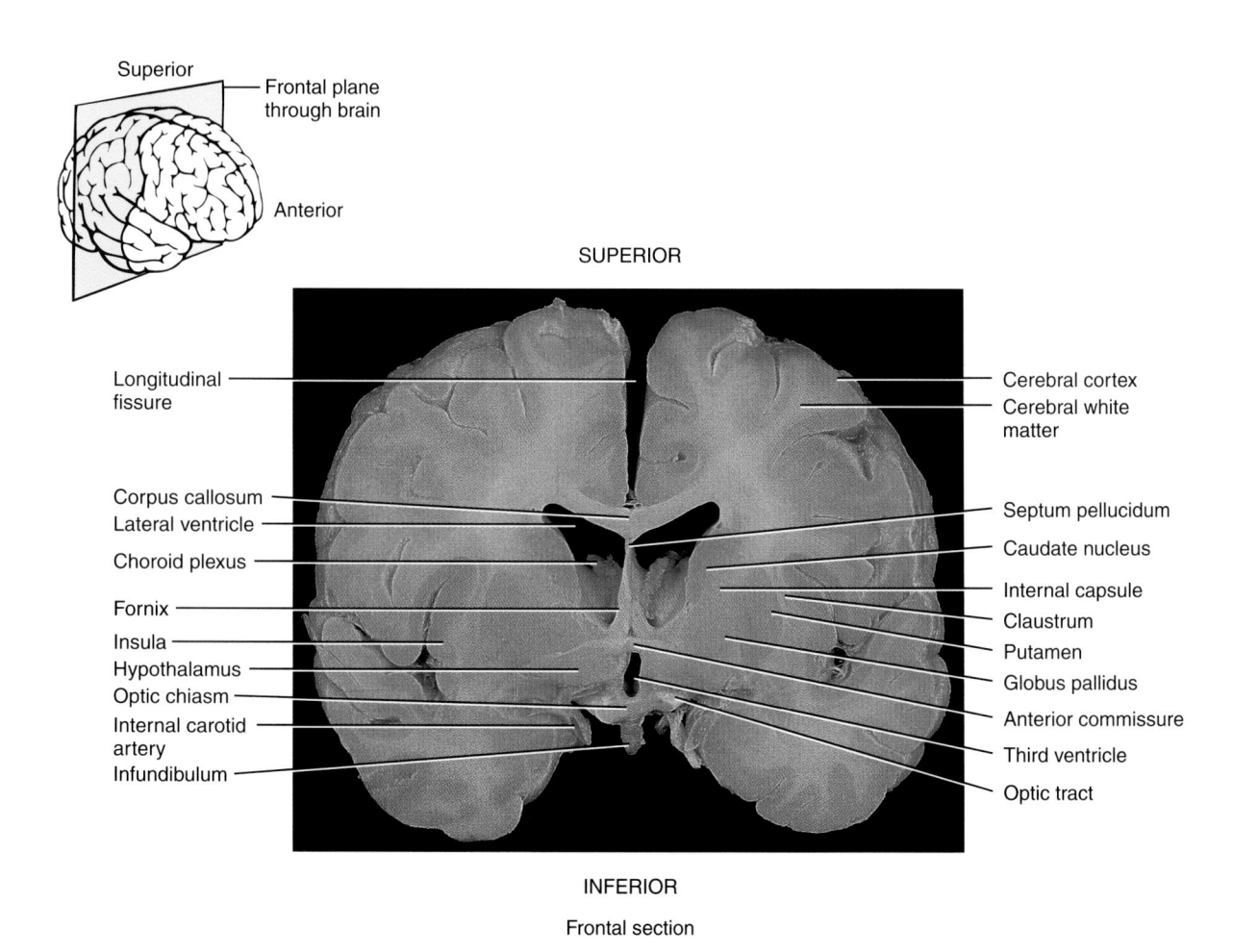

Superior

Frontal plane
through brain

Anterior

SUPERIOR

Longitudinal
fissure

Corpus callosum
Lateral ventricle
Choroid plexus
Fornix
Insula
Hypothalamus
Optic chiasm
Internal carotid
artery
Infundibulum

Cerebral cortex
Cerebral white
matter

Septum pellucidum
Caudate nucleus
Internal capsule
Claustrum
Putamen
Globus pallidus
Anterior commissure
Third ventricle
Optic tract

INFERIOR

Frontal section

FIGURE 8.18 | *Brain*

Inferior

ANTERIOR

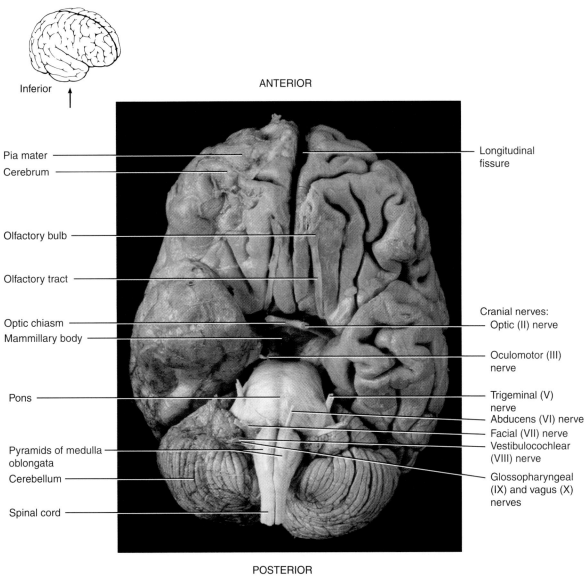

Pia mater

Cerebrum

Olfactory bulb

Olfactory tract

Optic chiasm

Mammillary body

Pons

Pyramids of medulla
oblongata

Cerebellum

Spinal cord

Longitudinal
fissure

Cranial nerves:
Optic (II) nerve

Oculomotor (III)
nerve

Trigeminal (V)
nerve
Abducens (VI) nerve
Facial (VII) nerve
Vestibulocochlear
(VIII) nerve
Glossopharyngeal
(IX) and vagus (X)
nerves

POSTERIOR

(a) Human brain

FIGURE 8.19 | *Brain*

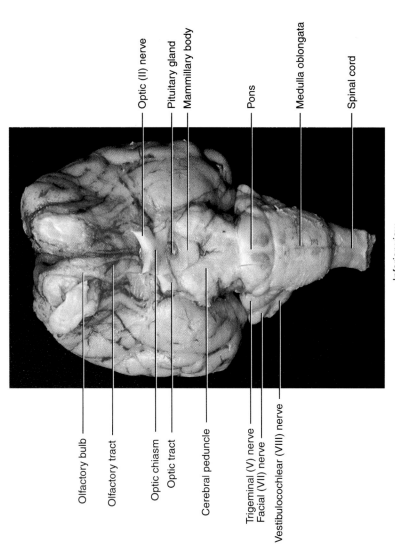

Olfactory bulb
Olfactory tract
Optic chiasm
Optic tract
Cerebral peduncle
Trigeminal (V) nerve
Facial (VII) nerve
Vestibulocochlear (VIII) nerve

Optic (II) nerve
Pituitary gland
Mammillary body
Pons
Medulla oblongata
Spinal cord

Inferior view
(b) Sheep brain

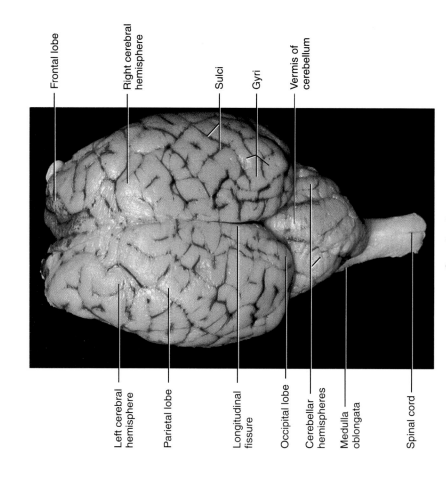

Frontal lobe
Right cerebral hemisphere
Sulci
Gyri
Vermis of cerebellum

Left cerebral hemisphere
Parietal lobe
Longitudinal fissure
Occipital lobe
Cerebellar hemispheres
Medulla oblongata
Spinal cord

Superior view
(c) Sheep brain

F I G U R E  8 . 1 9    *Brain, continued*

SUPERIOR

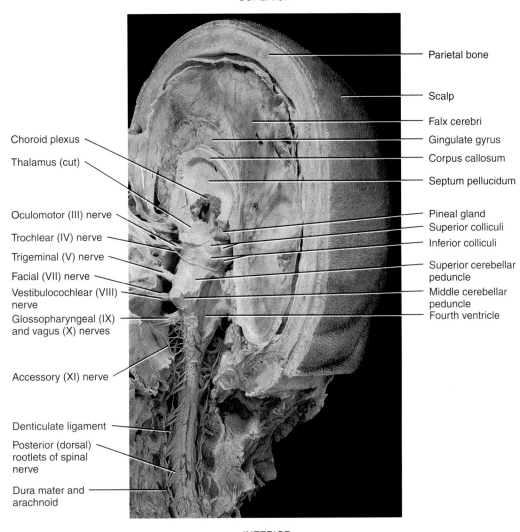

Parietal bone

Scalp

Falx cerebri

Gingulate gyrus

Corpus callosum

Septum pellucidum

Choroid plexus

Thalamus (cut)

Pineal gland
Superior colliculi
Inferior colliculi

Oculomotor (III) nerve

Trochlear (IV) nerve

Superior cerebellar peduncle

Trigeminal (V) nerve

Facial (VII) nerve

Middle cerebellar peduncle
Fourth ventricle

Vestibulocochlear (VIII) nerve

Glossopharyngeal (IX) and vagus (X) nerves

Accessory (XI) nerve

Denticulate ligament

Posterior (dorsal) rootlets of spinal nerve

Dura mater and arachnoid

INFERIOR

FIGURE 8.20 | *Brain (sagittal section) and spinal cord (posterior view)*

Superior

Transverse plane
through brain

Anterior

ANTERIOR

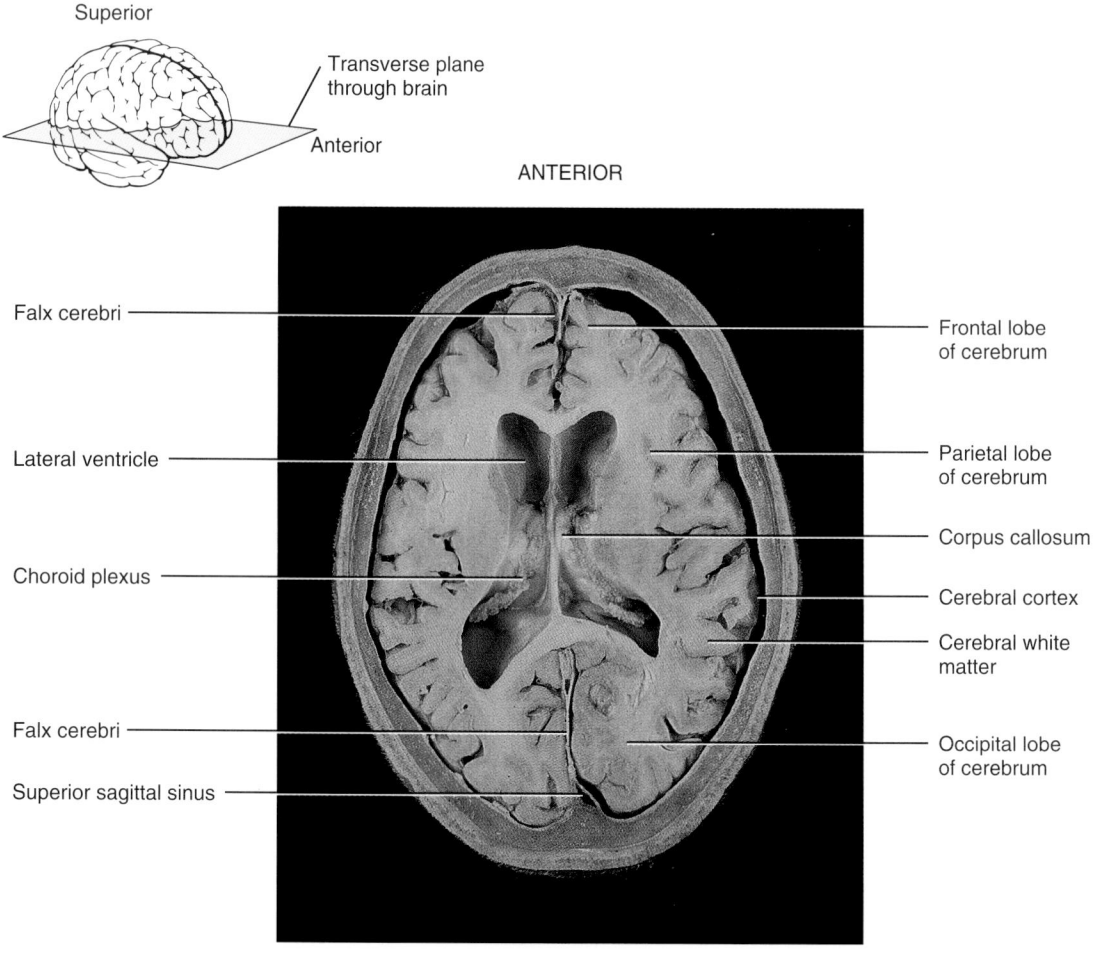

Falx cerebri

Lateral ventricle

Choroid plexus

Falx cerebri

Superior sagittal sinus

Frontal lobe
of cerebrum

Parietal lobe
of cerebrum

Corpus callosum

Cerebral cortex

Cerebral white
matter

Occipital lobe
of cerebrum

POSTERIOR

Transverse section

FIGURE 8.21 | *Brain*

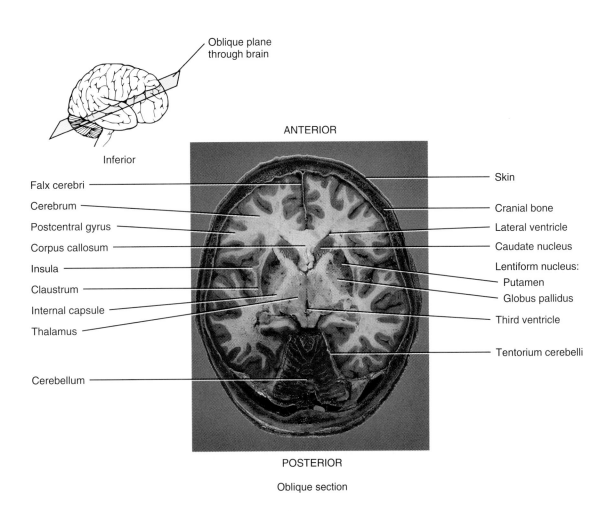

Oblique plane through brain

Inferior

ANTERIOR

Falx cerebri

Cerebrum

Postcentral gyrus

Corpus callosum

Insula

Claustrum

Internal capsule

Thalamus

Cerebellum

Skin

Cranial bone

Lateral ventricle

Caudate nucleus

Lentiform nucleus:

Putamen

Globus pallidus

Third ventricle

Tentorium cerebelli

POSTERIOR

Oblique section

FIGURE 8.22 | *Brain*

131

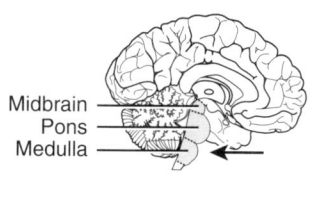

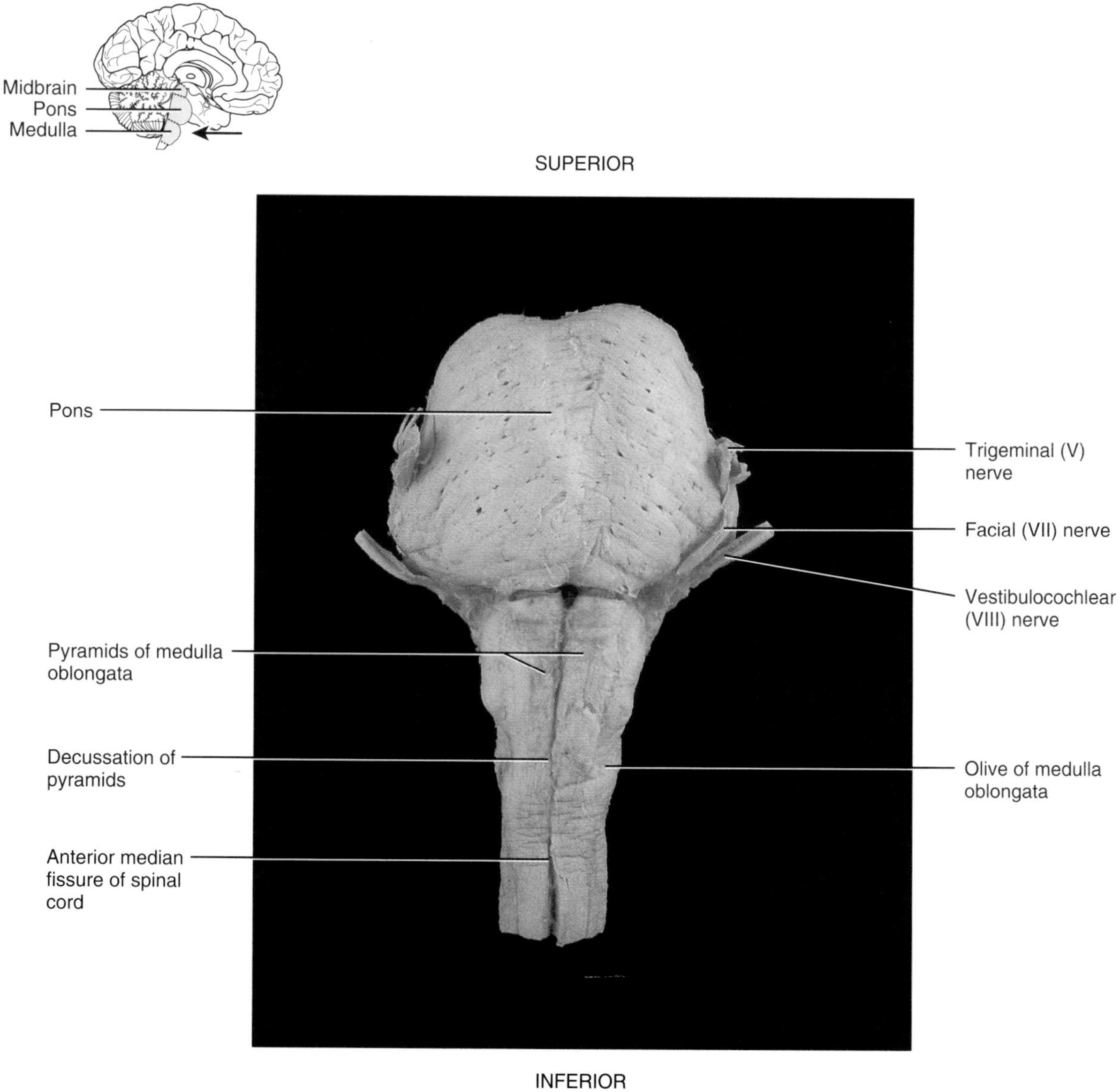

Midbrain
Pons
Medulla

SUPERIOR

Pons

Trigeminal (V)
nerve

Facial (VII) nerve

Vestibulocochlear
(VIII) nerve

Pyramids of medulla
oblongata

Decussation of
pyramids

Olive of medulla
oblongata

Anterior median
fissure of spinal
cord

INFERIOR

Anterior view
(a) Human brain

FIGURE 8.23 | *Brain stem*

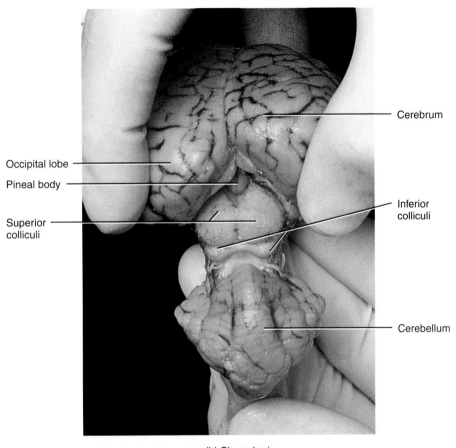

Cerebrum

Occipital lobe

Pineal body

Superior colliculi

Inferior colliculi

Cerebellum

(b) Sheep brain

133

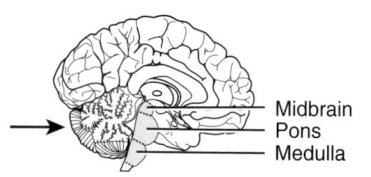

Midbrain
Pons
Medulla

SUPERIOR

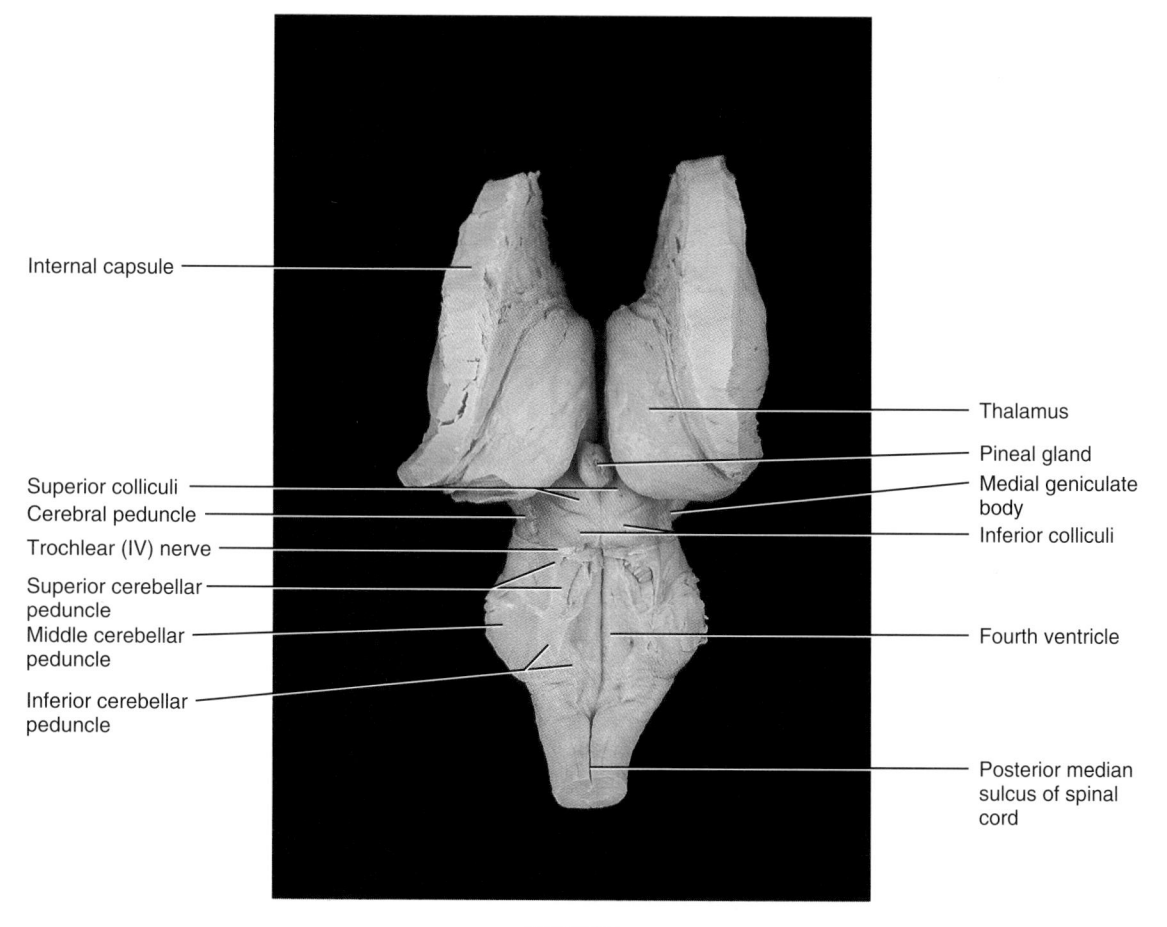

Internal capsule

Thalamus

Pineal gland

Superior colliculi
Cerebral peduncle
Trochlear (IV) nerve
Superior cerebellar
peduncle
Middle cerebellar
peduncle

Inferior cerebellar
peduncle

Medial geniculate
body
Inferior colliculi

Fourth ventricle

Posterior median
sulcus of spinal
cord

INFERIOR

Posterior view

F I G U R E  8 . 2 4  | *Brain stem*

Superior

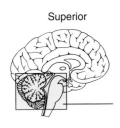

Midsaggital plane
through cerebellum
and brain stem

SUPERIOR

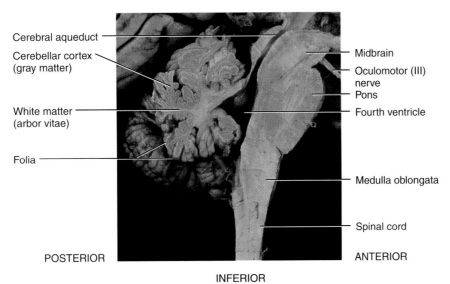

Cerebral aqueduct

Cerebellar cortex
(gray matter)

White matter
(arbor vitae)

Folia

Midbrain

Oculomotor (III)
nerve

Pons

Fourth ventricle

Medulla oblongata

Spinal cord

POSTERIOR

ANTERIOR

INFERIOR

Midsagittal section

FIGURE 8.25 | *Cerebellum and brain stem*

Neuroglia    Cell body

Processes

LM  260x

(a) Motor neuron

FIGURE 8.26 | *Histology of nervous tissue*

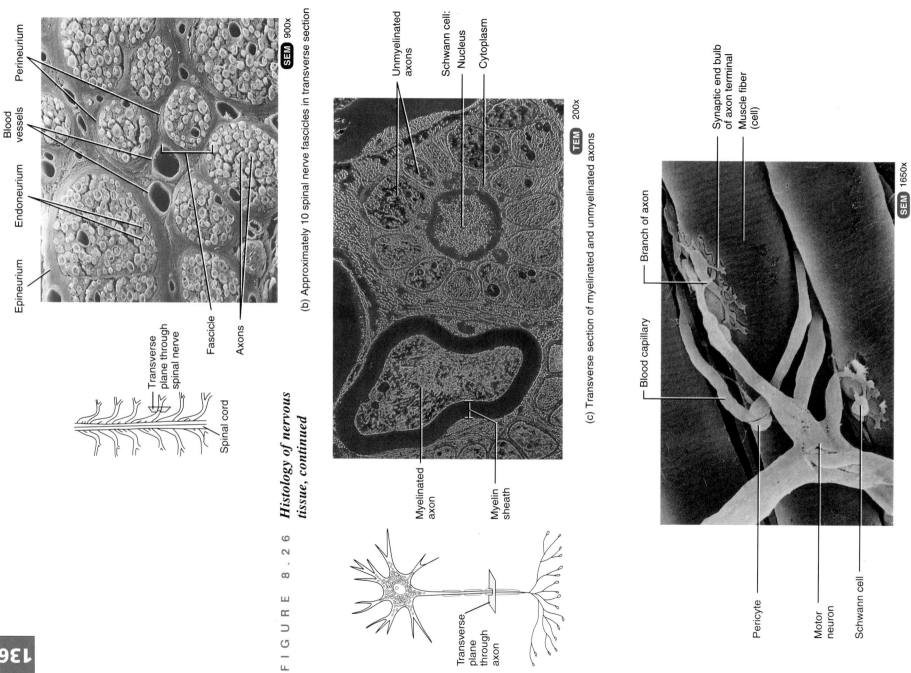

Perineurium

Blood vessels

Endoneurium

Epineurium

Fascicle

Axons

Transverse plane through spinal nerve

Spinal cord

**SEM** 900x

(b) Approximately 10 spinal nerve fascicles in transverse section

Unmyelinated axons

Schwann cell:
Nucleus
Cytoplasm

**TEM** 200x

Myelinated axon

Myelin sheath

Transverse plane through axon

(c) Transverse section of myelinated and unmyelinated axons

Synaptic end bulb of axon terminal
Muscle fiber (cell)

Branch of axon

Blood capillary

**SEM** 1650x

Pericyte

Motor neuron

Schwann cell

(d) Motor neuron and muscle fibers of a motor unit

FIGURE 8.26   *Histology of nervous tissue, continued*

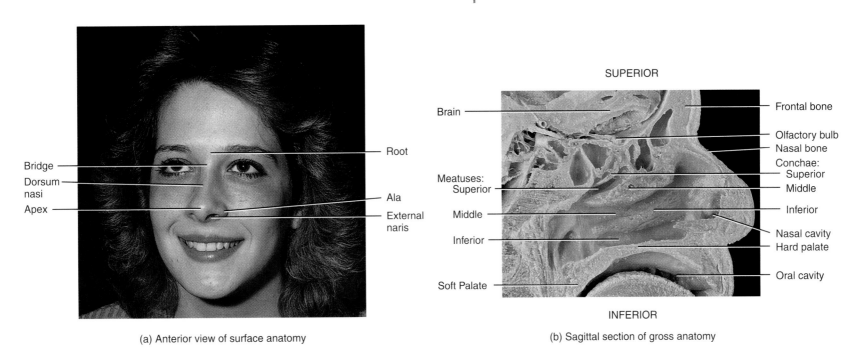

Bridge

Dorsum nasi

Apex

Root

Ala

External naris

(a) Anterior view of surface anatomy

SUPERIOR

Brain

Meatuses:
Superior

Middle

Inferior

Soft Palate

Frontal bone

Olfactory bulb

Nasal bone

Conchae:
Superior

Middle

Inferior

Nasal cavity

Hard palate

Oral cavity

INFERIOR

(b) Sagittal section of gross anatomy

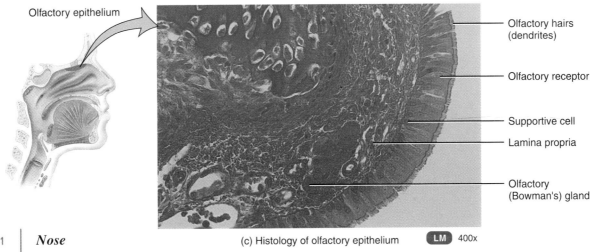

Olfactory epithelium

Olfactory hairs (dendrites)

Olfactory receptor

Supportive cell

Lamina propria

Olfactory (Bowman's) gland

FIGURE 9.1 | *Nose*

(c) Histology of olfactory epithelium   LM 400x

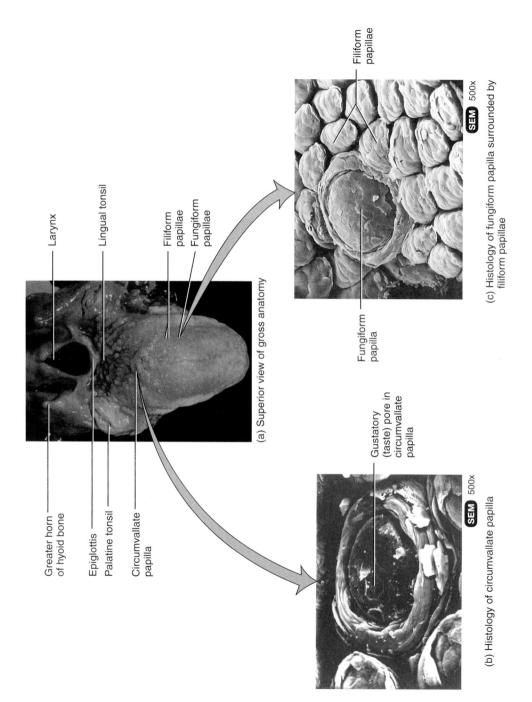

Larynx

Lingual tonsil

Filiform papillae
Fungiform papillae

Greater horn of hyoid bone

Epiglottis

Palatine tonsil

Circumvallate papilla

(a) Superior view of gross anatomy

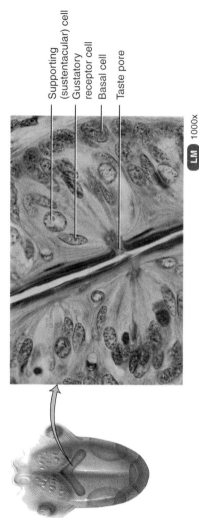

Filiform papillae

SEM 500x

Fungiform papilla

(c) Histology of fungiform papilla surrounded by filiform papillae

Gustatory (taste) pore in circumvallate papilla

SEM 500x

(b) Histology of circumvallate papilla

Supporting (sustentacular) cell
Gustatory receptor cell
Basal cell
Taste pore

LM 1000x

(d) Histology of a taste bud from a circumvallate papilla

FIGURE 9.2 | *Tongue*

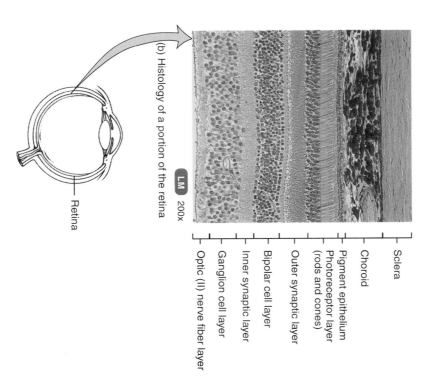

Retina

(b) Histology of a portion of the retina

LM 200x

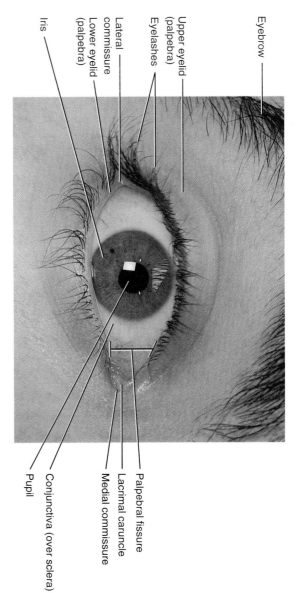

Optic (II) nerve fiber layer
Ganglion cell layer
Inner synaptic layer
Bipolar cell layer
Outer synaptic layer
Photoreceptor layer (rods and cones)
Pigment epithelium
Choroid
Sclera

(a) Anterior view of surface anatomy of right eye

Eyebrow

Upper eyelid (palpebra)
Eyelashes
Lateral commissure
Lower eyelid (palpebra)

Iris

Pupil
Conjunctiva (over sclera)
Medial commissure
Lacrimal caruncle
Palpebral fissure

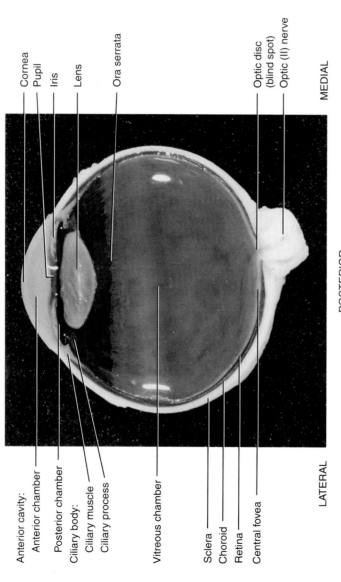

ANTERIOR

Cornea
Pupil
Iris
Lens

Ora serrata

Optic disc
(blind spot)
Optic (II) nerve

MEDIAL

Anterior cavity:
Anterior chamber
Posterior chamber
Ciliary body:
Ciliary muscle
Ciliary process

Vitreous chamber

Sclera
Choroid
Retina
Central fovea

LATERAL

POSTERIOR

(c) Superior view of transverse section of gross anatomy of human eye

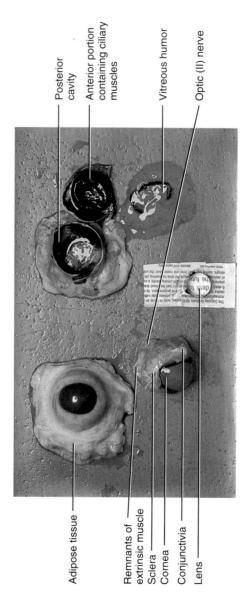

Posterior cavity

Anterior portion containing ciliary muscles

Vitreous humor

Optic (II) nerve

Adipose tissue

Remnants of extrinsic muscle
Sclera
Cornea
Conjunctivia
Lens

(d) Gross anatomy of the cow eye

FIGURE 9.3 *Eye, continued*

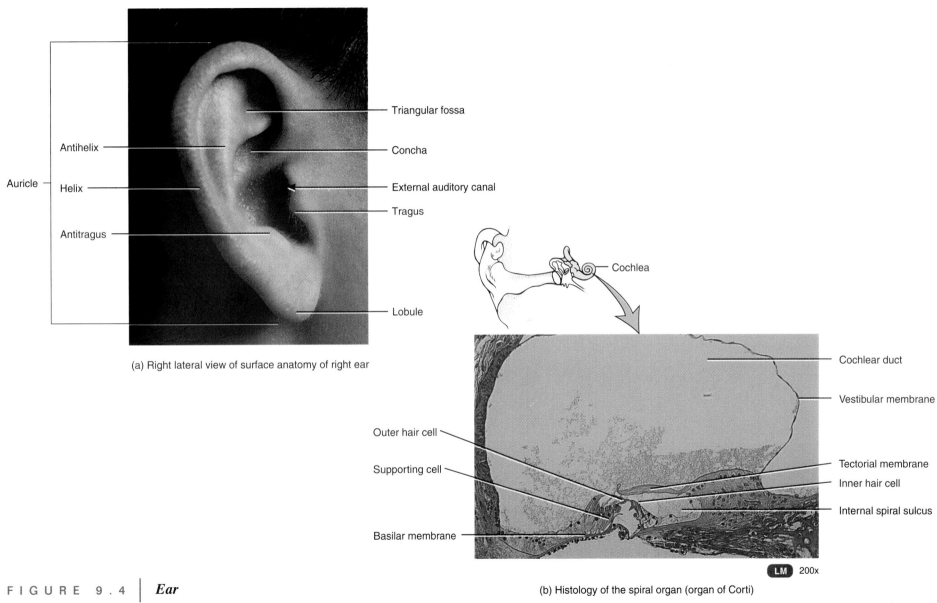

Auricle

Antihelix

Helix

Antitragus

Triangular fossa

Concha

External auditory canal

Tragus

Lobule

(a) Right lateral view of surface anatomy of right ear

Cochlea

Cochlear duct

Vestibular membrane

Outer hair cell

Supporting cell

Basilar membrane

Tectorial membrane

Inner hair cell

Internal spiral sulcus

LM 200x

(b) Histology of the spiral organ (organ of Corti)

FIGURE 9.4 | *Ear*

141

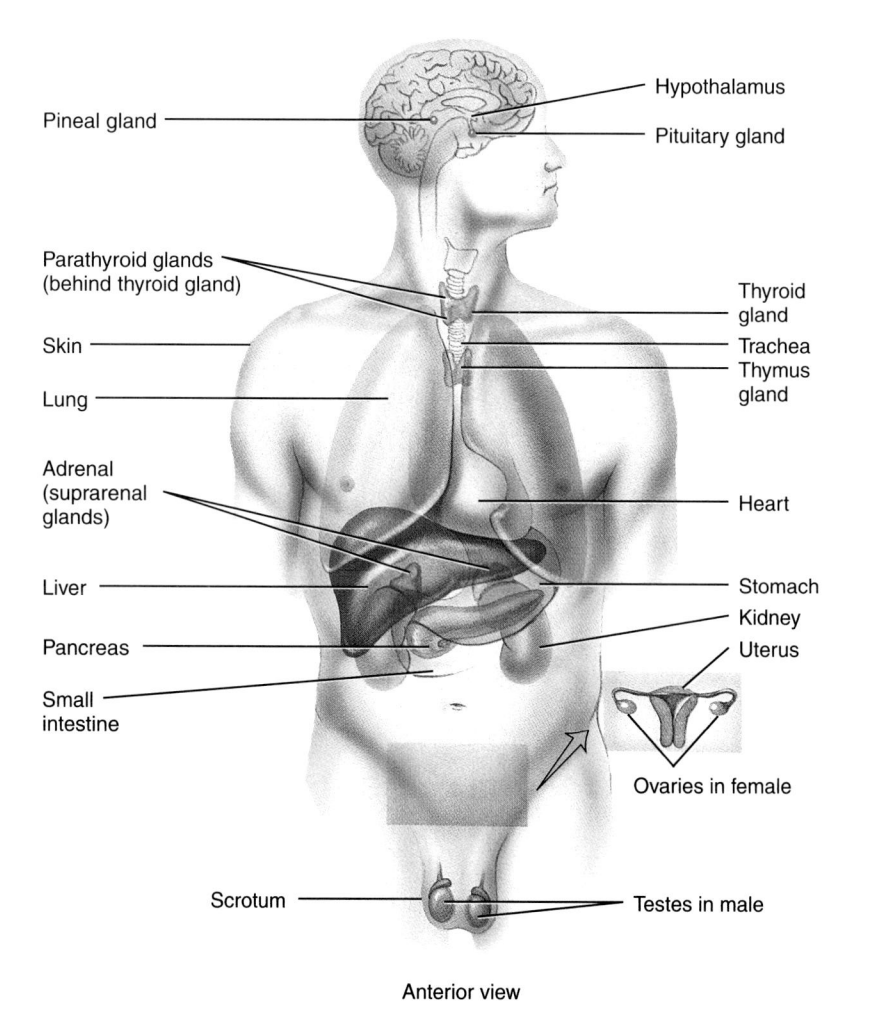

Pineal gland

Hypothalamus

Pituitary gland

Parathyroid glands
(behind thyroid gland)

Thyroid gland

Skin

Trachea

Thymus gland

Lung

Adrenal
(suprarenal glands)

Heart

Liver

Stomach

Kidney

Pancreas

Uterus

Small intestine

Ovaries in female

Scrotum

Testes in male

Anterior view

FIGURE 10.1 | *Endocrine glands, organs containing endocrine tissue, and associated structures*

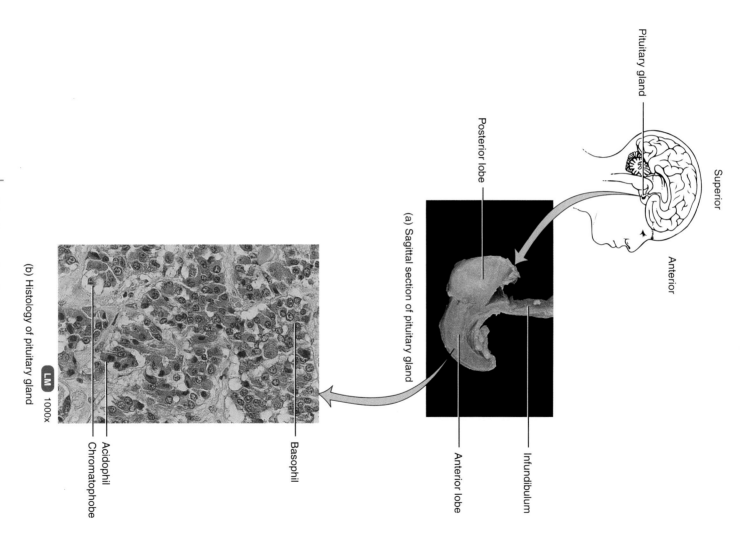

Superior

Anterior

Pituitary gland

Posterior lobe

Anterior lobe

Infundibulum

(a) Sagittal section of pituitary gland

**LM** 1000x

Acidophil
Chromatophobe

Basophil

(b) Histology of pituitary gland

F I G U R E  1 0 . 2 | *Pituitary gland*

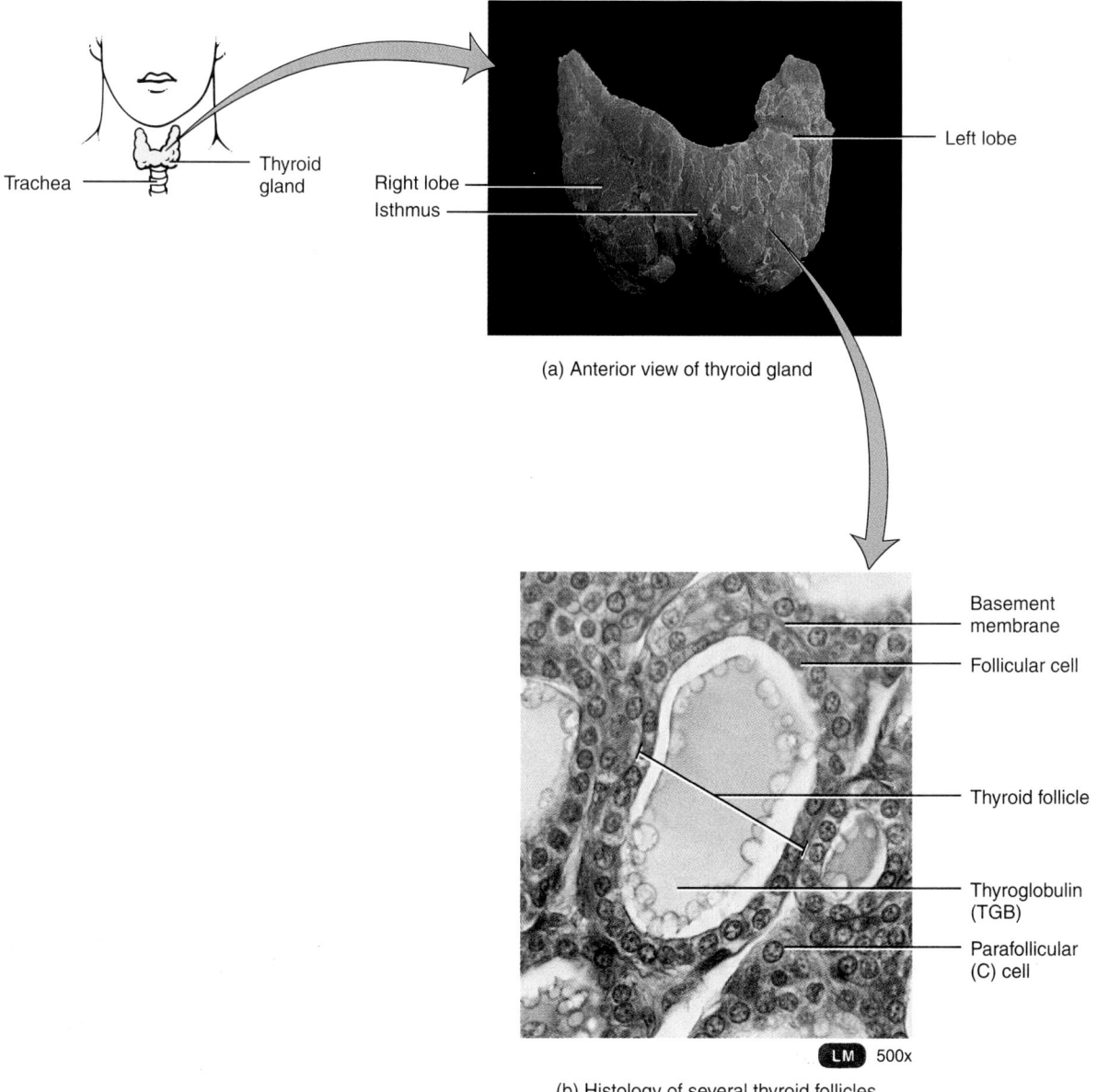

Trachea

Thyroid gland

Left lobe

Right lobe

Isthmus

(a) Anterior view of thyroid gland

Basement membrane

Follicular cell

Thyroid follicle

Thyroglobulin (TGB)

Parafollicular (C) cell

LM 500x

(b) Histology of several thyroid follicles

FIGURE 10.3 | *Thyroid gland*

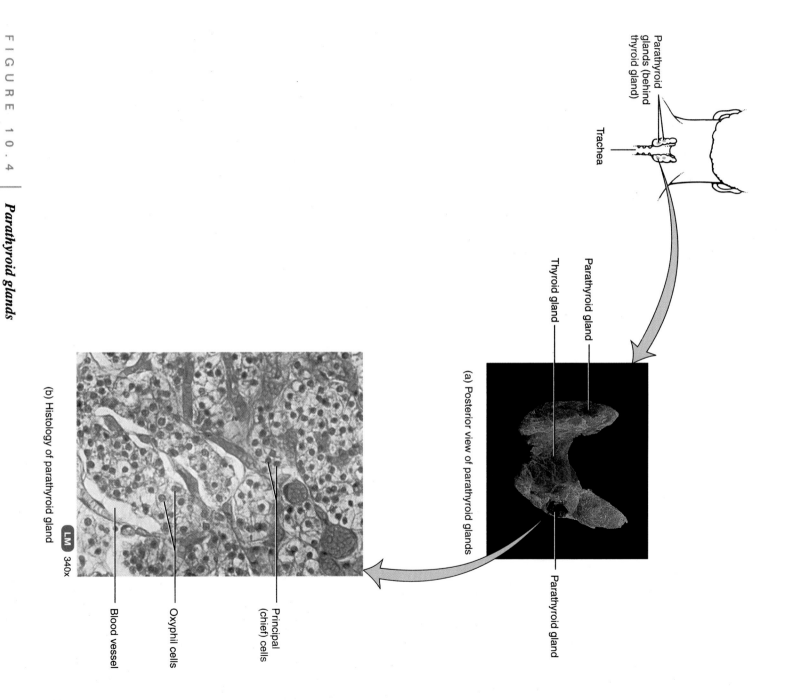

FIGURE 10.4 | *Parathyroid glands*

Parathyroid glands (behind thyroid gland)

Trachea

(a) Posterior view of parathyroid glands

Parathyroid gland

Thyroid gland

Parathyroid gland

(b) Histology of parathyroid gland

LM 340x

Blood vessel

Oxyphil cells

Principal (chief) cells

FIGURE 10.5 *Adrenal (suprarenal) glands*

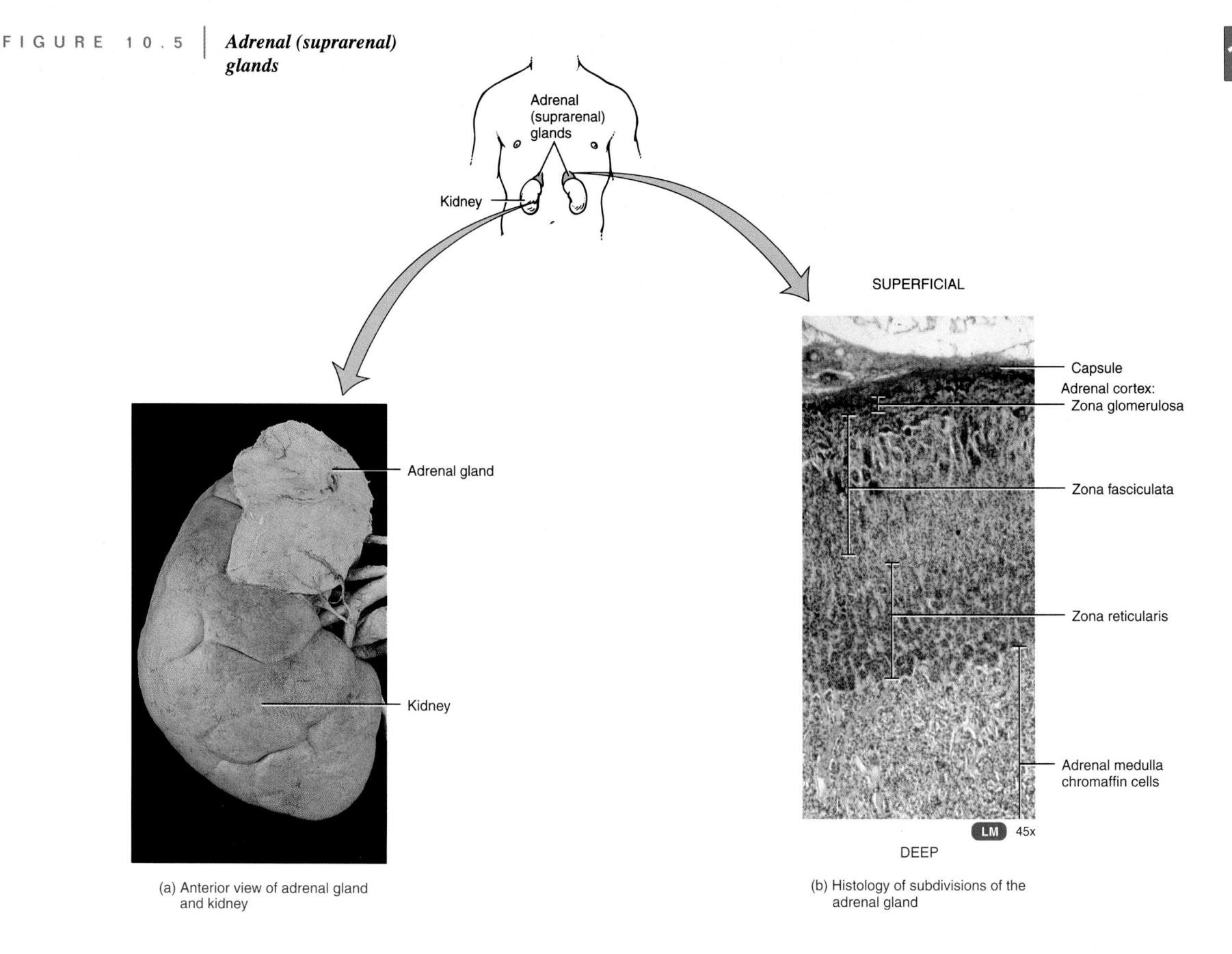

(a) Anterior view of adrenal gland and kidney

(b) Histology of subdivisions of the adrenal gland

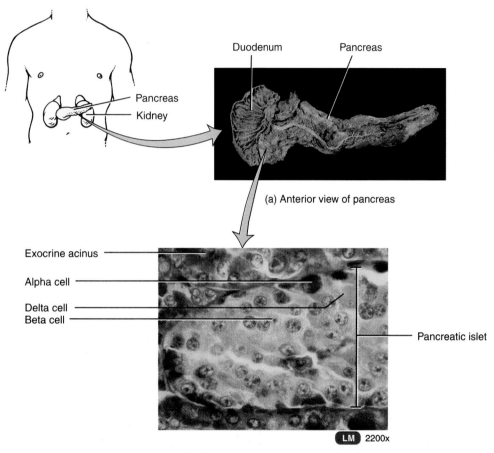

Duodenum          Pancreas

Pancreas
Kidney

(a) Anterior view of pancreas

Exocrine acinus

Alpha cell

Delta cell
Beta cell

Pancreatic islet

LM 2200x

(b) Histology of pancreatic islet (islet of
Langerhans) and surrounding acini

FIGURE 10.6 | *Pancreas*

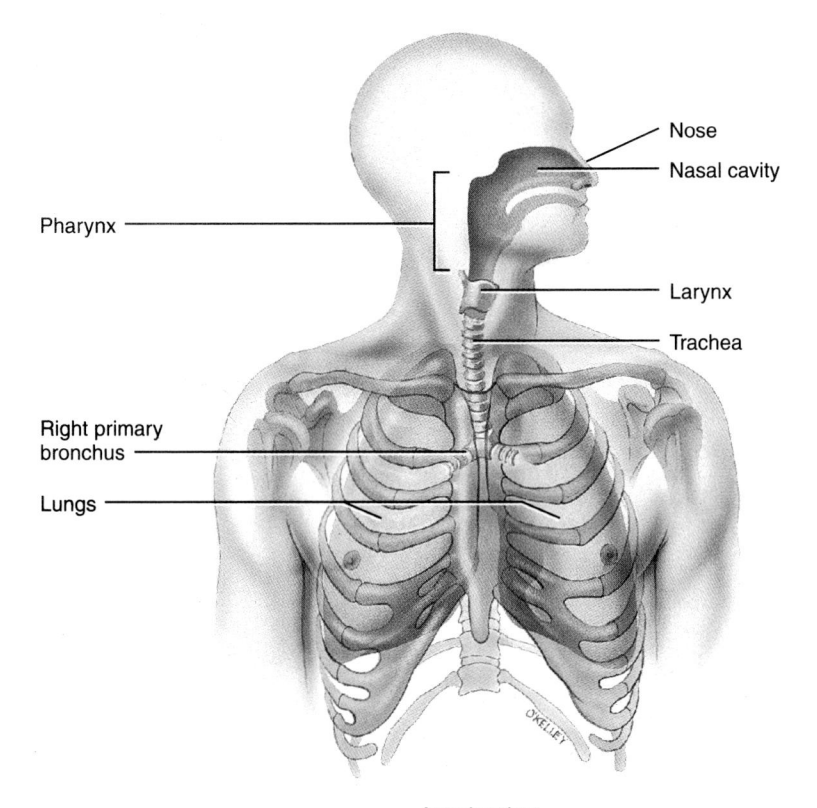

Nose

Nasal cavity

Pharynx

Larynx

Trachea

Right primary
bronchus

Lungs

Anterior view

FIGURE 11.1 | *Respiratory System*

SUPERIOR

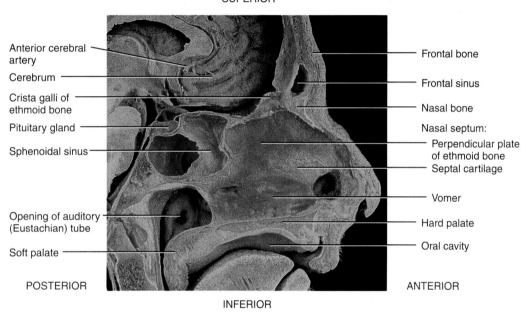

Anterior cerebral artery

Cerebrum

Crista galli of ethmoid bone

Pituitary gland

Sphenoidal sinus

Opening of auditory (Eustachian) tube

Soft palate

Frontal bone

Frontal sinus

Nasal bone

Nasal septum:

Perpendicular plate of ethmoid bone

Septal cartilage

Vomer

Hard palate

Oral cavity

POSTERIOR

ANTERIOR

INFERIOR

Right lateral view

FIGURE 11.2 | *Nasal septum*

149

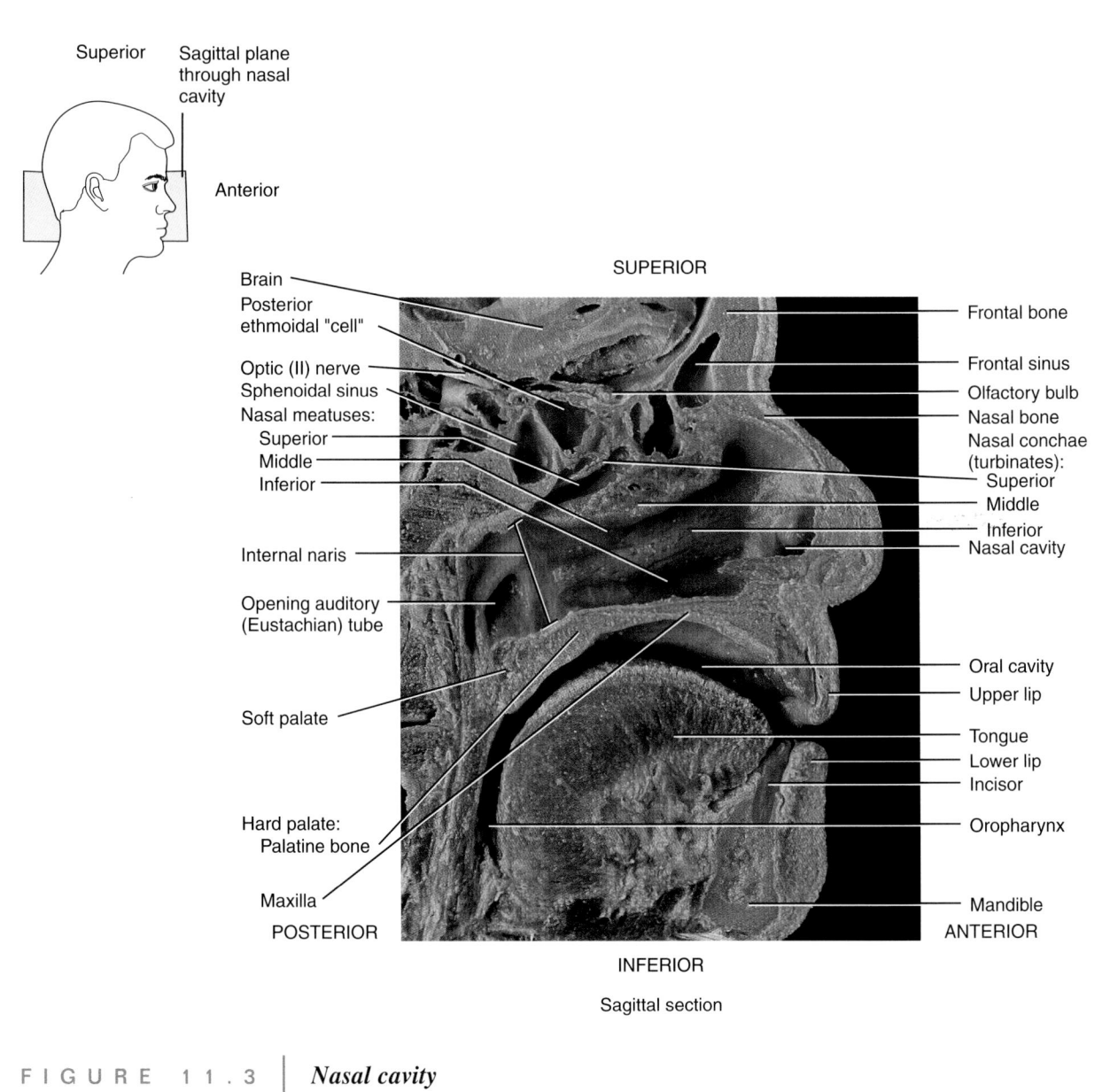

Superior    Sagittal plane
            through nasal
            cavity

Anterior

SUPERIOR

Brain

Posterior
ethmoidal "cell"

Optic (II) nerve

Sphenoidal sinus

Nasal meatuses:
    Superior
    Middle
    Inferior

Internal naris

Opening auditory
(Eustachian) tube

Soft palate

Hard palate:
    Palatine bone

Maxilla

POSTERIOR

Frontal bone

Frontal sinus

Olfactory bulb

Nasal bone

Nasal conchae
(turbinates):
    Superior
    Middle
    Inferior
Nasal cavity

Oral cavity

Upper lip

Tongue
Lower lip
Incisor

Oropharynx

Mandible

ANTERIOR

INFERIOR

Sagittal section

FIGURE  11.3  │  *Nasal cavity*

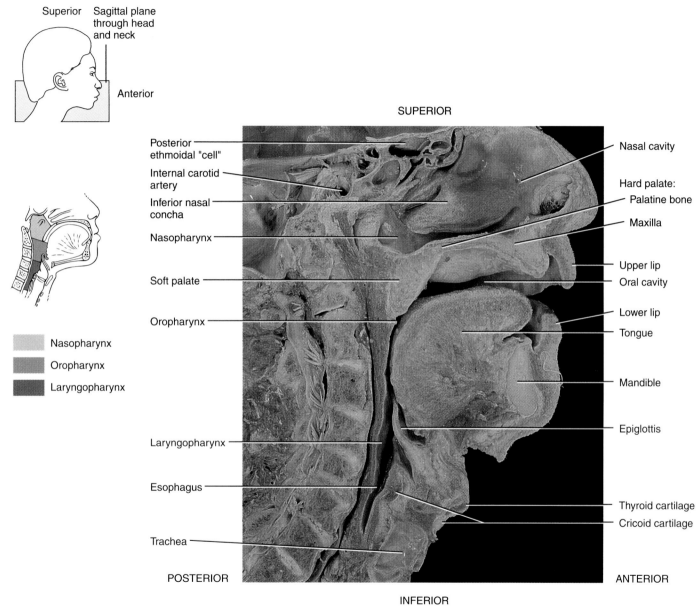

Superior   Sagittal plane
            through head
            and neck

Anterior

Nasopharynx
Oropharynx
Laryngopharynx

SUPERIOR

Posterior ethmoidal "cell"

Internal carotid artery

Inferior nasal concha

Nasopharynx

Soft palate

Oropharynx

Laryngopharynx

Esophagus

Trachea

POSTERIOR

Nasal cavity

Hard palate:
    Palatine bone

Maxilla

Upper lip
Oral cavity

Lower lip
Tongue

Mandible

Epiglottis

Thyroid cartilage
Cricoid cartilage

ANTERIOR

INFERIOR

Sagittal section

FIGURE 11.4 | *Subdivisions of pharynx*

151

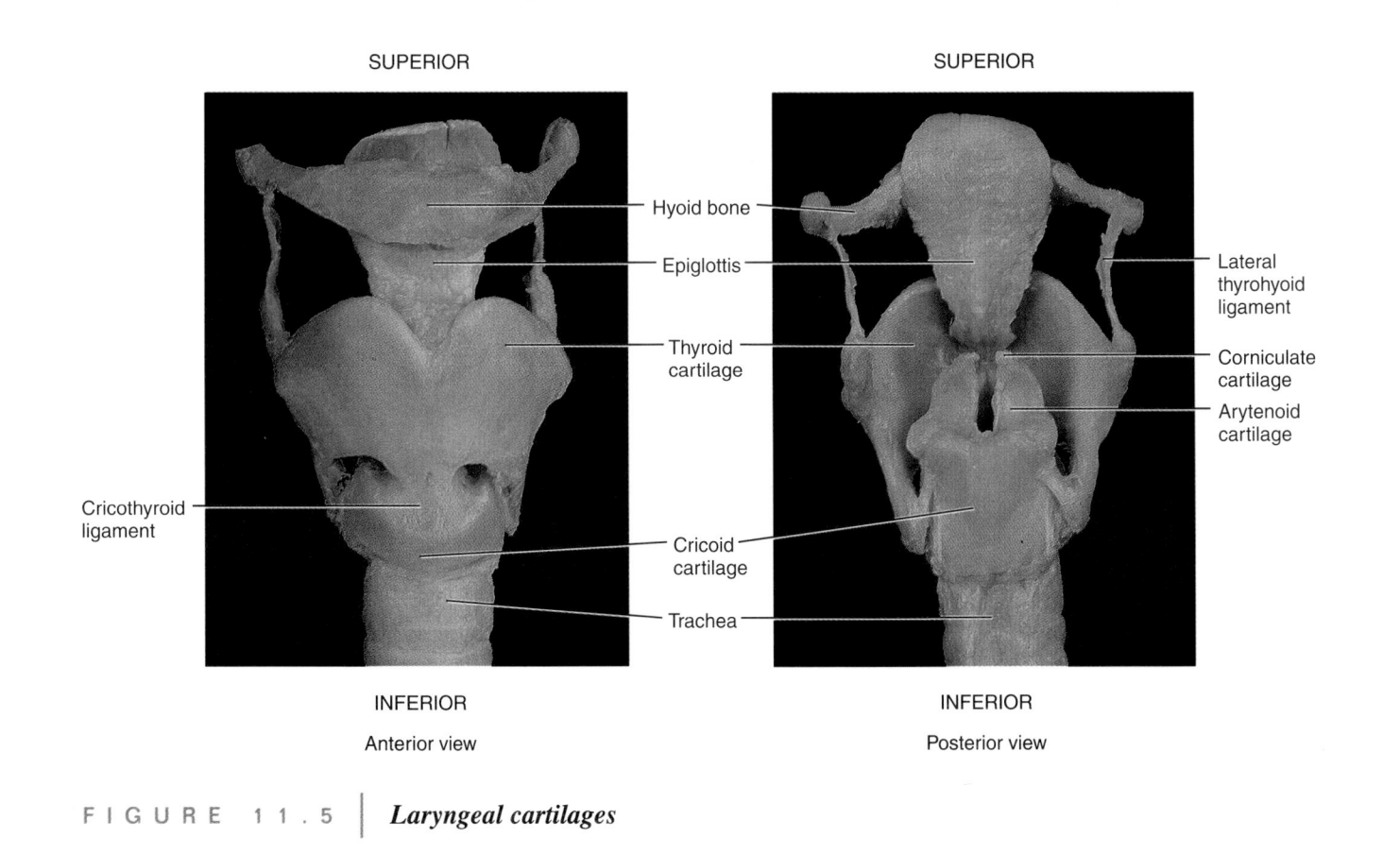

SUPERIOR

SUPERIOR

Hyoid bone

Epiglottis

Lateral thyrohyoid ligament

Thyroid cartilage

Corniculate cartilage

Arytenoid cartilage

Cricothyroid ligament

Cricoid cartilage

Trachea

INFERIOR

INFERIOR

Anterior view

Posterior view

FIGURE 11.5 | *Laryngeal cartilages*

SUPERIOR

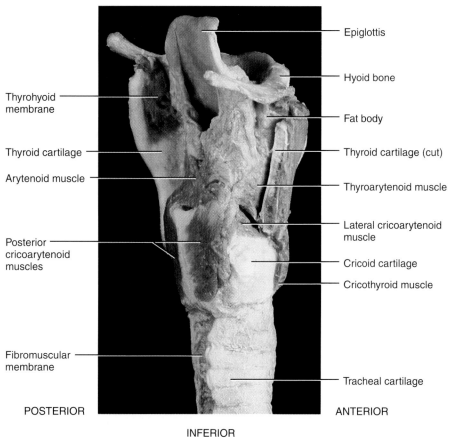

Epiglottis

Hyoid bone

Thyrohyoid membrane

Fat body

Thyroid cartilage

Thyroid cartilage (cut)

Arytenoid muscle

Thyroarytenoid muscle

Lateral cricoarytenoid muscle

Posterior cricoarytenoid muscles

Cricoid cartilage

Cricothyroid muscle

Fibromuscular membrane

Tracheal cartilage

POSTERIOR

ANTERIOR

INFERIOR

Right posterolateral view

FIGURE 11.6 | *Larynx*

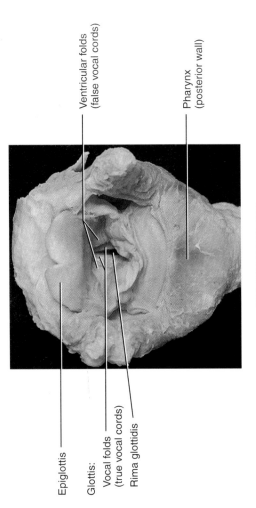

SUPEROANTERIOR

Epiglottis

Glottis:

Ventricular folds
(false vocal cords)

Vocal folds
(true vocal cords)

Rima glottidis

Pharynx
(posterior wall)

INFEROPOSTERIOR
Superolateral view

FIGURE 11.7 | *Larynx*

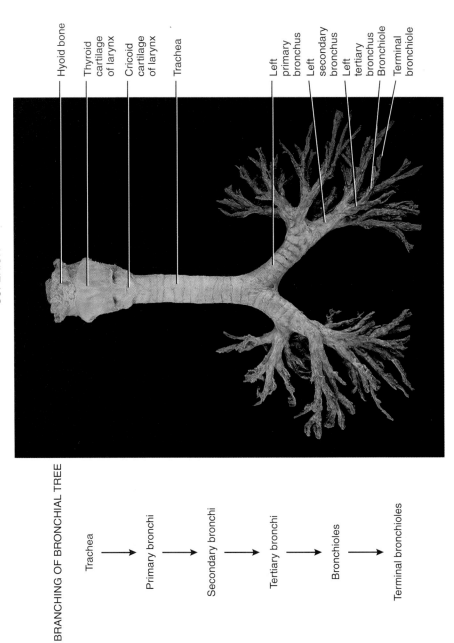

SUPERIOR

Hyoid bone

Thyroid
cartilage
of larynx

Cricoid
cartilage
of larynx

Trachea

Left
primary
bronchus

Left
secondary
bronchus

Left
tertiary
bronchus

Bronchiole

Terminal
bronchiole

INFERIOR
Anterior view

BRANCHING OF BRONCHIAL TREE

Trachea → Primary bronchi → Secondary bronchi → Tertiary bronchi → Bronchioles → Terminal bronchioles

FIGURE 11.8 | *Bronchial tree*

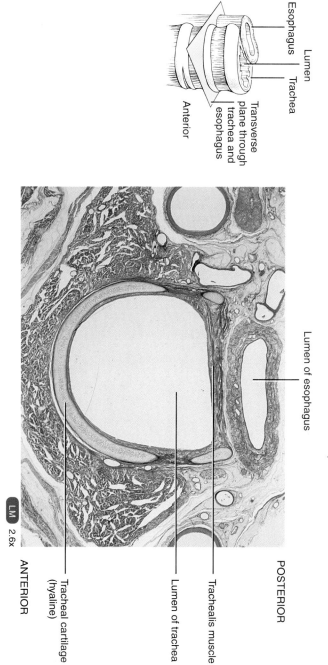

Esophagus

Lumen

Trachea

Transverse
plane through
trachea and
esophagus

Anterior

LM 2.6x

ANTERIOR

Tracheal cartilage
(hyaline)

Lumen of trachea

Trachealis muscle

POSTERIOR

Lumen of esophagus

(a) Transverse section of the trachea in relation to the esophagus

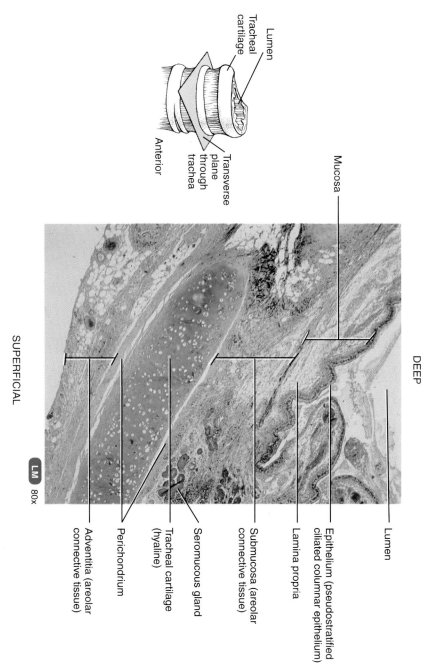

Lumen
Tracheal
cartilage

Anterior

Transverse
plane
through
trachea

Mucosa

SUPERFICIAL

DEEP

LM 80x

Adventitia (areolar
connective tissue)

Perichondrium

Tracheal cartilage
(hyaline)

Seromucous gland

Submucosa (areolar
connective tissue)

Lamina propria

Epithelium (pseudostratified
ciliated columnar epithelium)

Lumen

(b) Transverse section of part of the tracheal wall

155

Mucus in goblet cell    Cilia    Lumen

Nucleus of ciliated
columnar cell

Nucleus of goblet cell

Nucleus of basal cell

Basement membrane

Lamina propria
(areolar connective tissue)

Epithelium
(pseudostratified
ciliated columnar
epithelium)

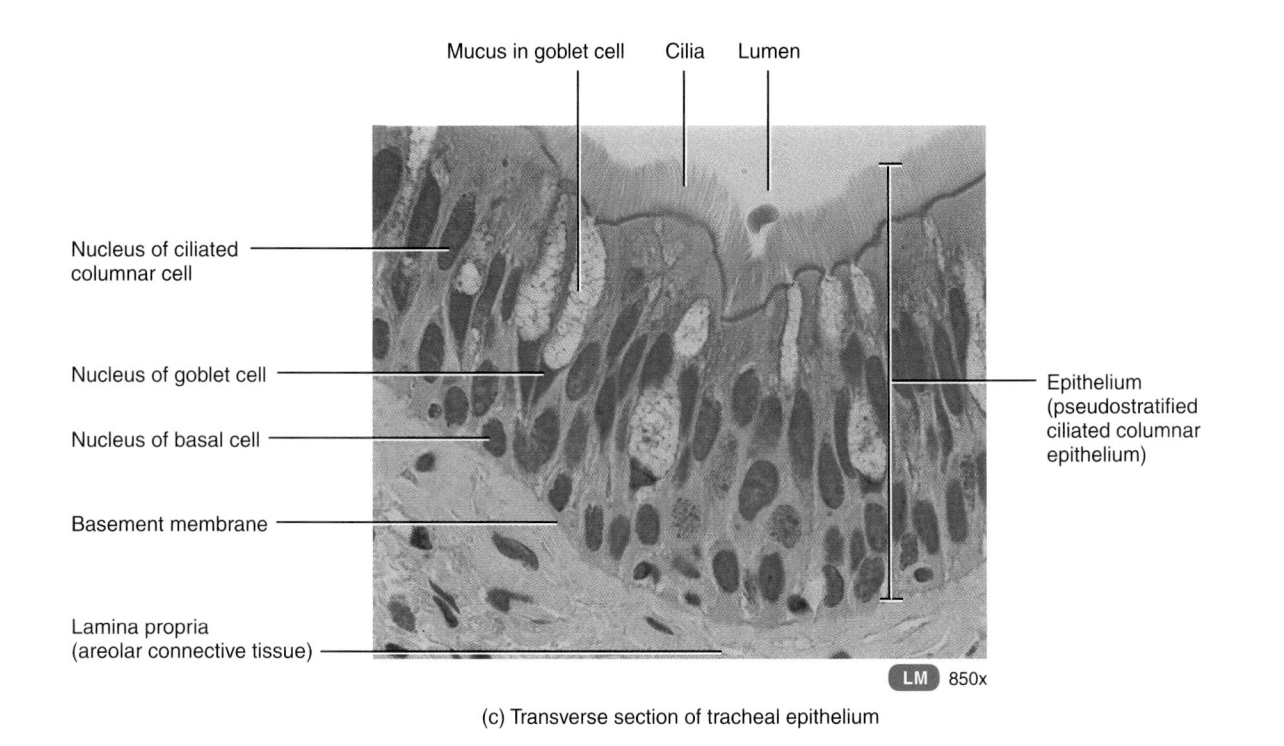

LM 850x

(c) Transverse section of tracheal epithelium

FIGURE  11.9  *Histology of the trachea,*
*continued*

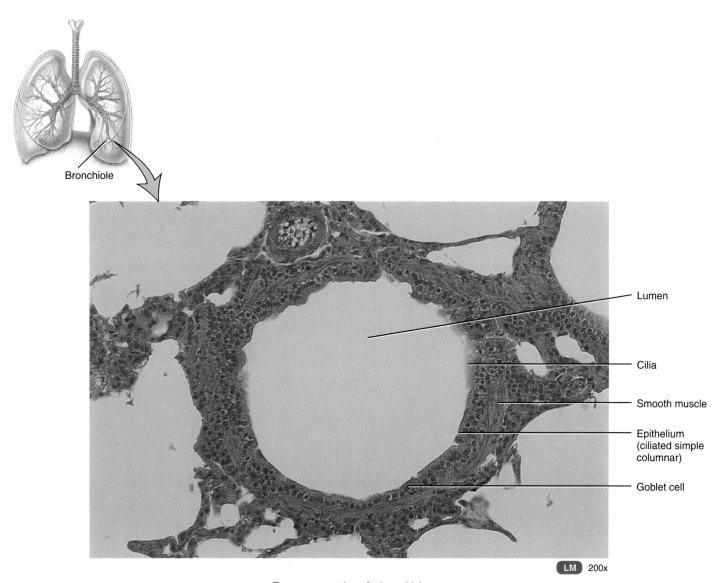

Bronchiole

Lumen

Cilia

Smooth muscle

Epithelium
(ciliated simple
columnar)

Goblet cell

LM 200x

Transverse section of a bronchiole

FIGURE 11.10 | *Histology of a bronchiole*

SUPERIOR

Trachea   Esophagus

Body of thoracic vertebra

Vertebral artery

Rib

Left subclavian artery

Left brachiocephalic vein

Left common carotid artery

Right brachiocephalic vein

Left vagus (X) nerve

Brachiocephalic trunk

Superior vena cava

Arch of aorta

Pulmonary trunk

Ascending aorta

Left lung

Right lung

Rib

Oblique fissure

Diaphragm

Fibrous pericardium around heart

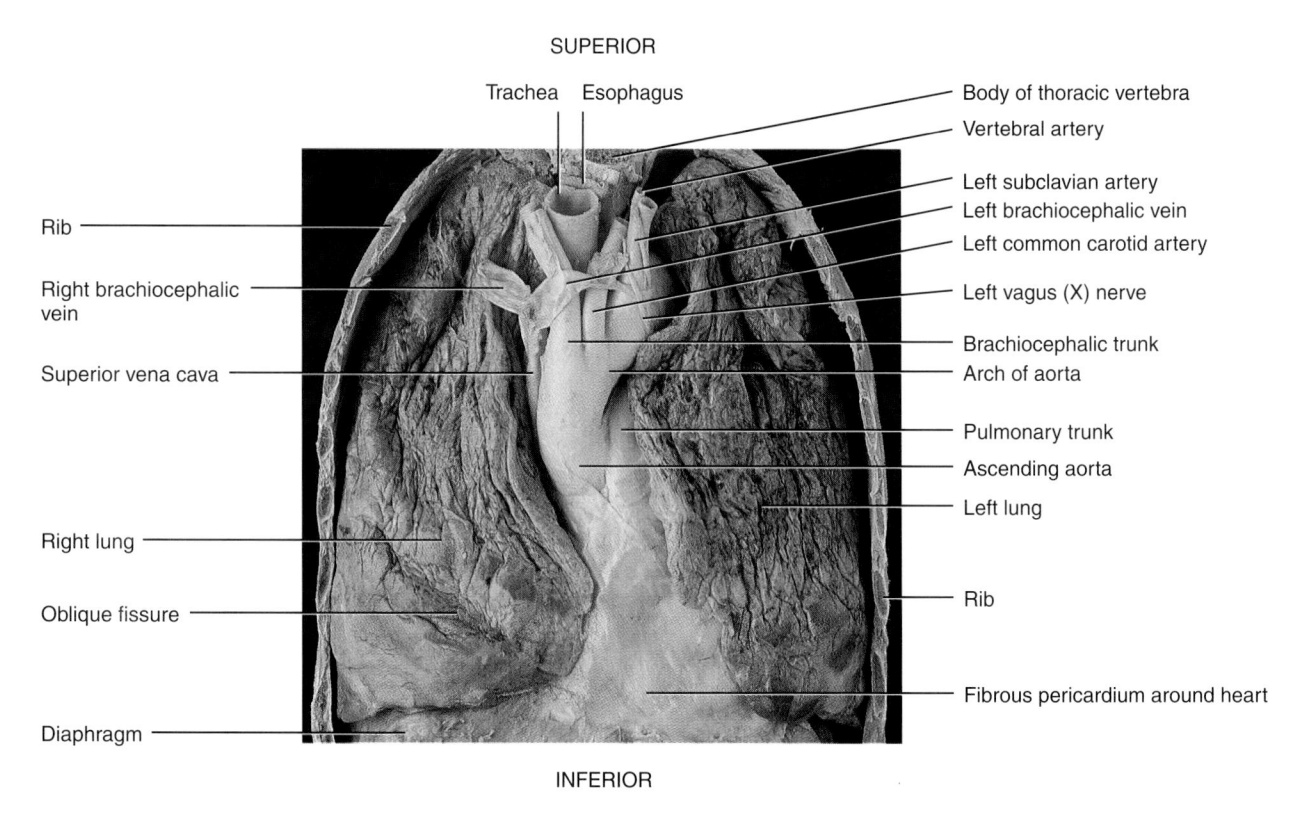

INFERIOR

Anterior view of the lungs after removal of the anterolateral
thoracic wall and parietal pleura

FIGURE 11.11 | *Lungs*

FIGURE 11.12 | *Lungs*

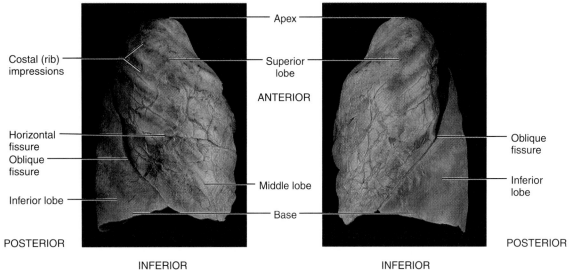

SUPERIOR

Apex

Costal (rib) impressions

Superior lobe

ANTERIOR

Horizontal fissure

Oblique fissure

Oblique fissure

Inferior lobe

Inferior lobe

Middle lobe

Base

POSTERIOR

POSTERIOR

INFERIOR

INFERIOR

Right lung, lateral view

Left lung, lateral view

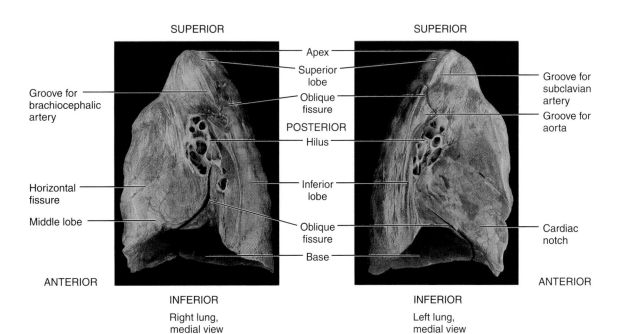

SUPERIOR

SUPERIOR

Apex

Superior lobe

Groove for brachiocephalic artery

Oblique fissure

Groove for subclavian artery

POSTERIOR

Groove for aorta

Hilus

Inferior lobe

Horizontal fissure

Middle lobe

Oblique fissure

Cardiac notch

Base

ANTERIOR

ANTERIOR

INFERIOR

INFERIOR

Right lung, medial view

Left lung, medial view

159

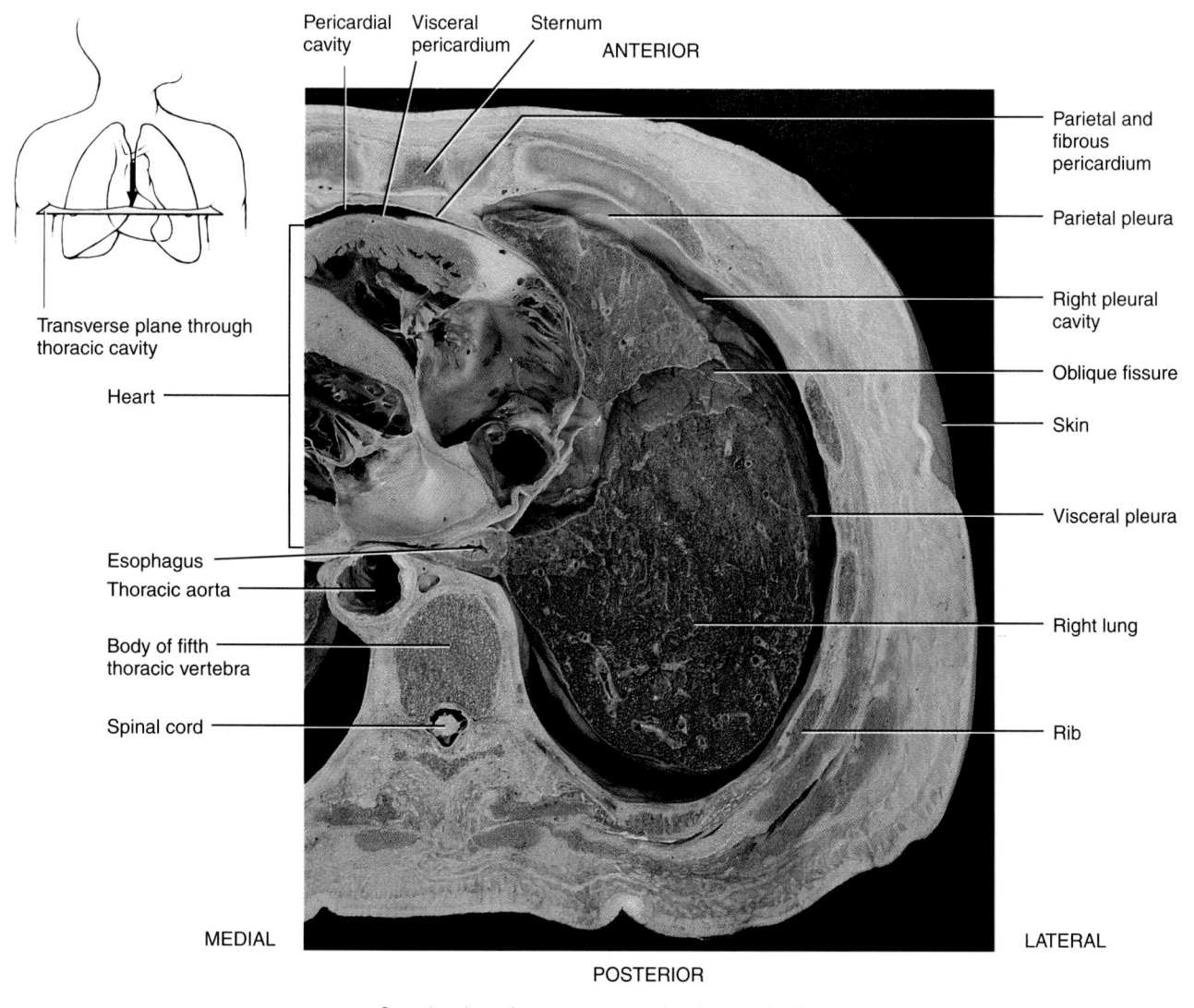

Pericardial cavity

Visceral pericardium

Sternum

ANTERIOR

Parietal and fibrous pericardium

Parietal pleura

Right pleural cavity

Oblique fissure

Skin

Visceral pleura

Right lung

Rib

Transverse plane through thoracic cavity

Heart

Esophagus

Thoracic aorta

Body of fifth thoracic vertebra

Spinal cord

MEDIAL

POSTERIOR

LATERAL

Superior view of a transverse section through the thoracic cavity

FIGURE 11.13 | *Lungs*

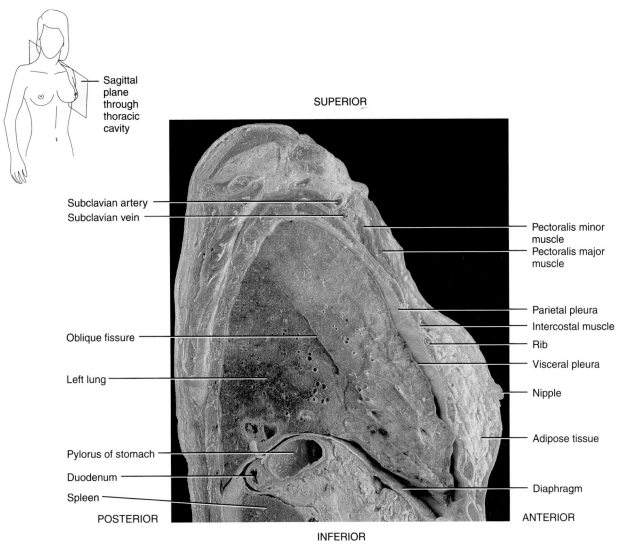

Sagittal plane through thoracic cavity

SUPERIOR

Subclavian artery

Subclavian vein

Pectoralis minor muscle

Pectoralis major muscle

Parietal pleura

Intercostal muscle

Oblique fissure

Rib

Visceral pleura

Left lung

Nipple

Adipose tissue

Pylorus of stomach

Duodenum

Spleen

Diaphragm

POSTERIOR

ANTERIOR

INFERIOR

Sagittal section

FIGURE 11.14 | *Lungs*

FIGURE 11.15 | *Histology of the lungs*

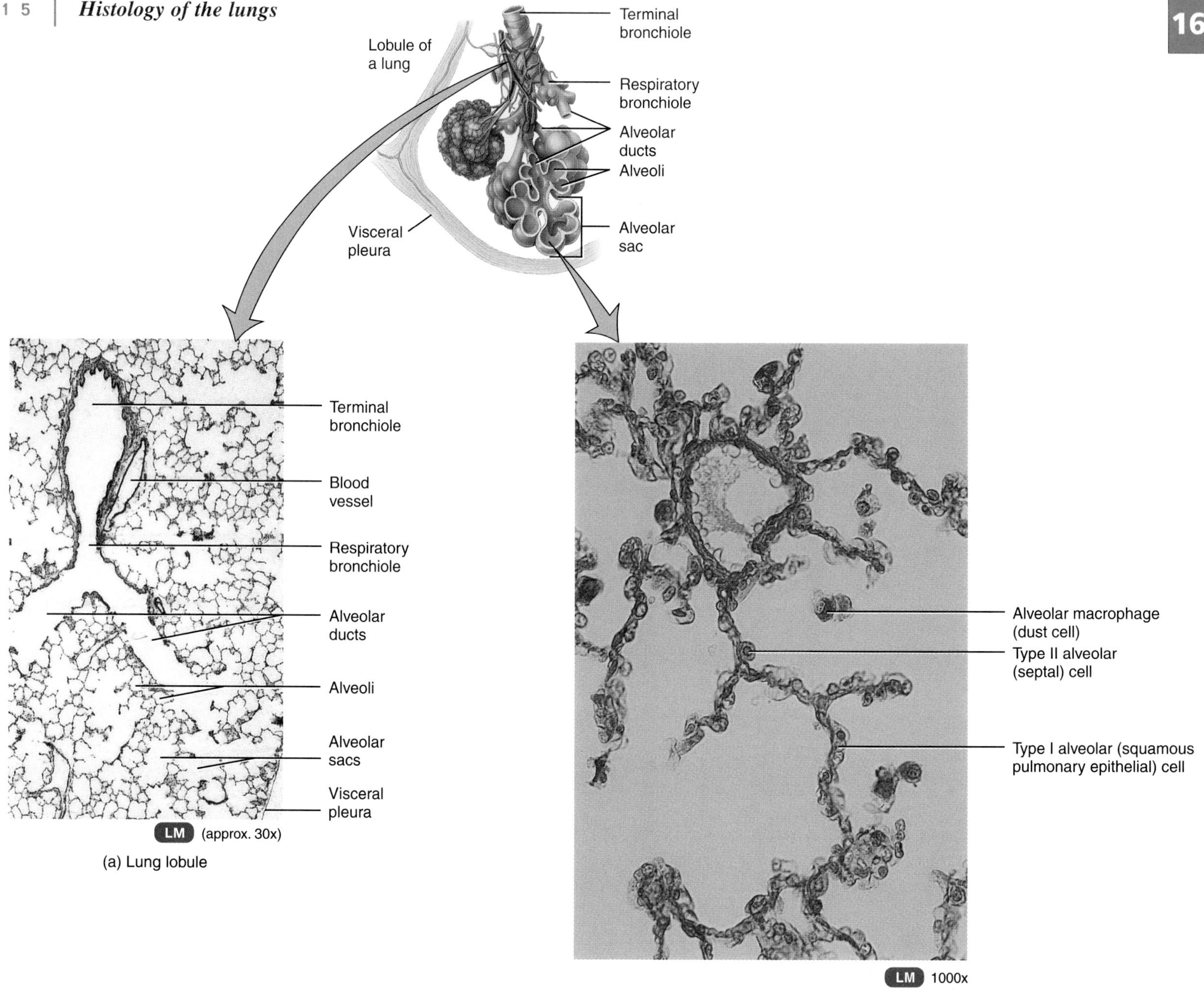

162

Lobule of
a lung

Terminal
bronchiole

Respiratory
bronchiole

Alveolar
ducts

Alveoli

Alveolar
sac

Visceral
pleura

Terminal
bronchiole

Blood
vessel

Respiratory
bronchiole

Alveolar
ducts

Alveoli

Alveolar
sacs

Visceral
pleura

LM (approx. 30x)

(a) Lung lobule

Alveolar macrophage
(dust cell)

Type II alveolar
(septal) cell

Type I alveolar (squamous
pulmonary epithelial) cell

LM 1000x

(b) Alveoli

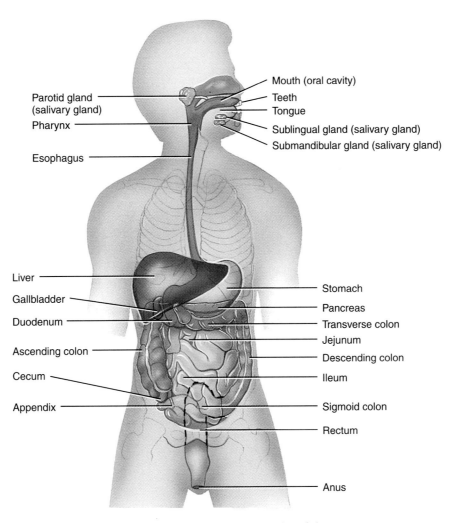

Parotid gland
(salivary gland)

Pharynx

Esophagus

Mouth (oral cavity)

Teeth

Tongue

Sublingual gland (salivary gland)

Submandibular gland (salivary gland)

Liver

Gallbladder

Duodenum

Ascending colon

Cecum

Appendix

Stomach

Pancreas

Transverse colon

Jejunum

Descending colon

Ileum

Sigmoid colon

Rectum

Anus

Right lateral view of head and neck and anterior view of chest,
abdomen, and pelvis

FIGURE 12.1 | *Digestive system*

SUPERIOR

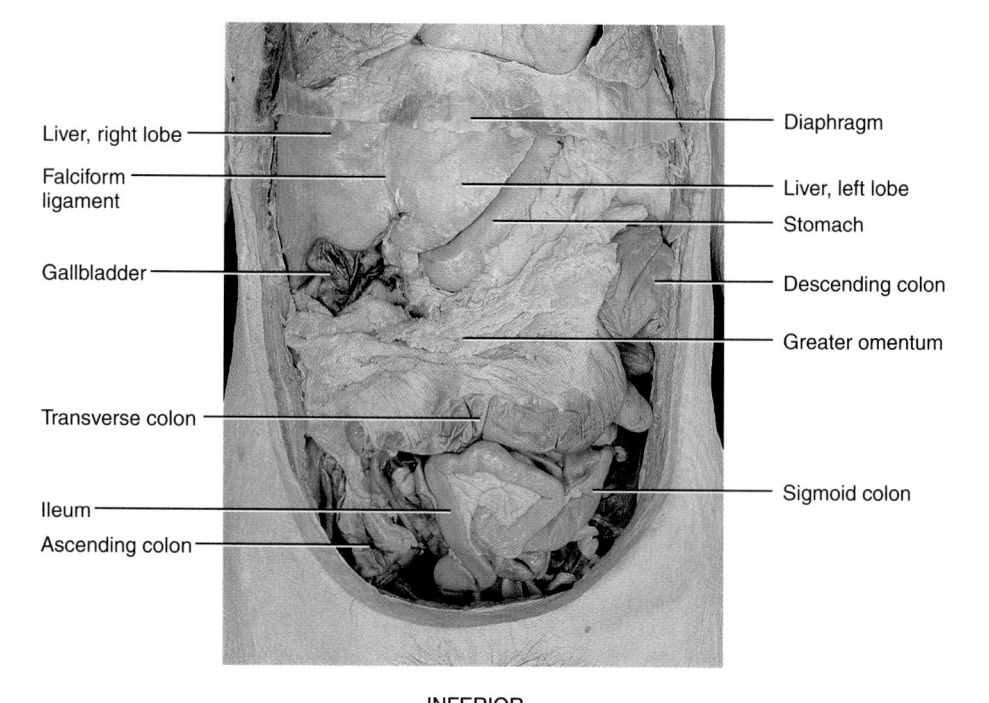

Liver, right lobe

Falciform
ligament

Gallbladder

Transverse colon

Ileum

Ascending colon

Diaphragm

Liver, left lobe

Stomach

Descending colon

Greater omentum

Sigmoid colon

INFERIOR

Anterior view

FIGURE 1 2 . 2 | *Digestive organs of the abdominal cavity*

SUPERIOR

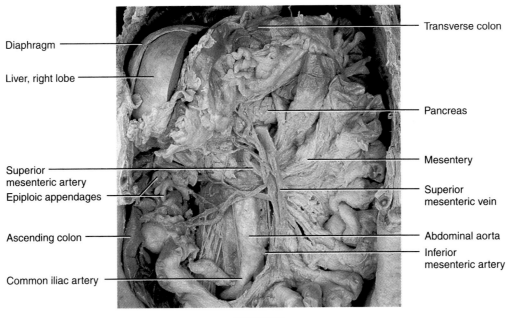

Diaphragm

Liver, right lobe

Superior
mesenteric artery

Epiploic appendages

Ascending colon

Common iliac artery

Transverse colon

Pancreas

Mesentery

Superior
mesenteric vein

Abdominal aorta

Inferior
mesenteric artery

INFERIOR

Anterior view

FIGURE 12.3 | *Digestive organs of the abdominal cavity*

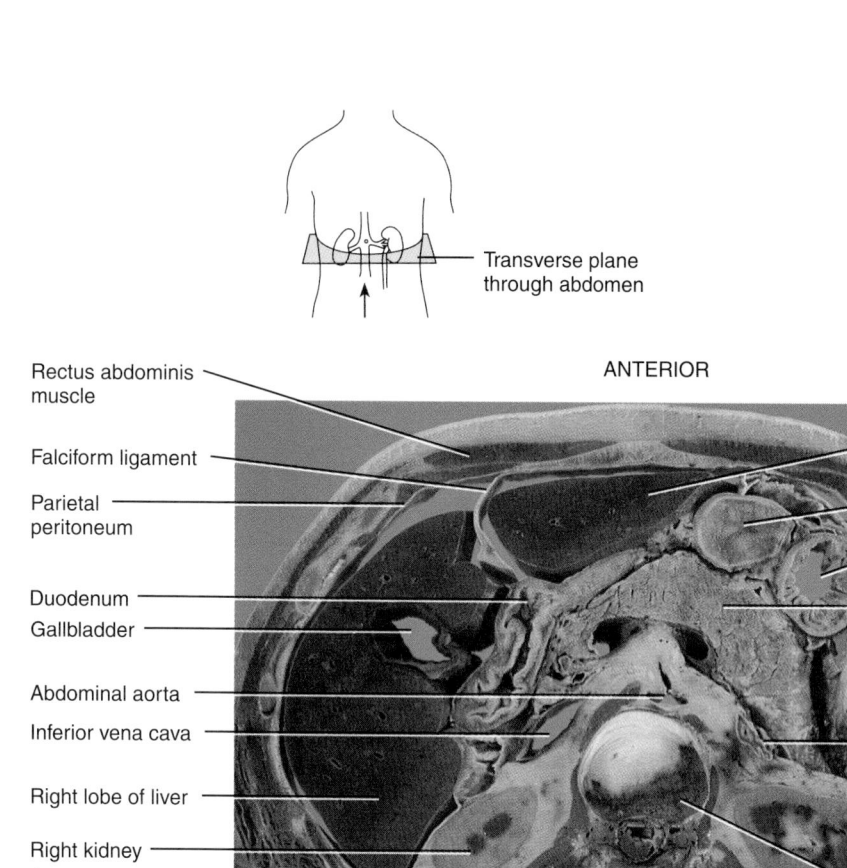

Transverse plane
through abdomen

ANTERIOR

Rectus abdominis
muscle

Falciform ligament

Parietal
peritoneum

Duodenum

Gallbladder

Abdominal aorta

Inferior vena cava

Right lobe of liver

Right kidney

Adipose capsule
(perirenal fat)

Erector spinae
muscle

Left lobe of liver

Stomach

Rib

Pancreas

Transverse
colon

Left adrenal
(suprarenal)
gland

Descending
colon

Left kidney

Body of T12 vetebra

POSTERIOR

Inferior view of a transverse section through the abdomen

FIGURE  12.4  | *Digestive organs of the*
*abdominal cavity*

FIGURE 12.5 | *Digestive organs of the upper abdominal cavity*

Sagittal plane through upper abdomen

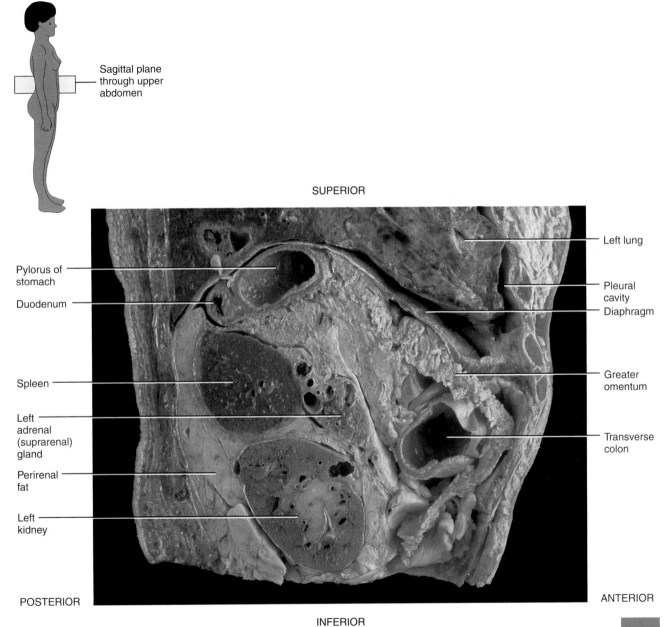

SUPERIOR

Pylorus of stomach

Duodenum

Spleen

Left adrenal (suprarenal) gland

Perirenal fat

Left kidney

POSTERIOR

Left lung

Pleural cavity

Diaphragm

Greater omentum

Transverse colon

ANTERIOR

INFERIOR

Sagittal section

POSTERIOR

SUPERIOR

ANTERIOR

Masseter muscle

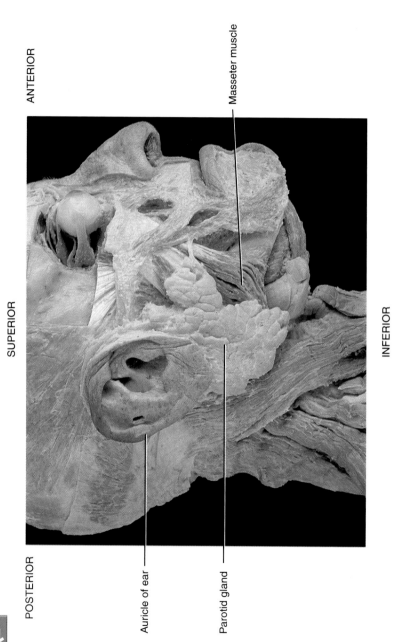

Auricle of ear

Parotid gland

INFERIOR

(a) Right lateral view of gross anatomy

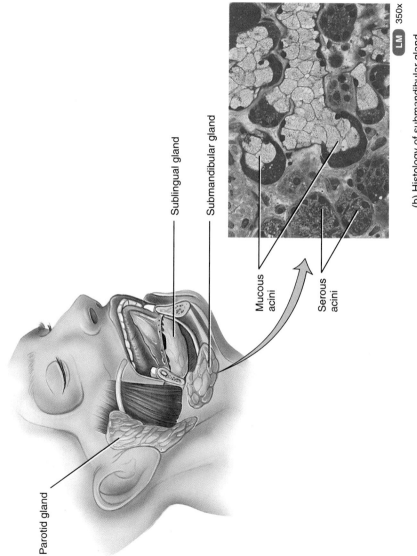

Parotid gland

Sublingual gland

Submandibular gland

Mucous acini

Serous acini

LM 350x

(b) Histology of submandibular gland

FIGURE 12.6 | *Salivary glands*

FIGURE 12.7 | *Teeth*

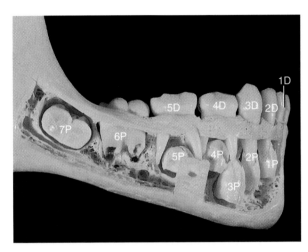

Right lateral view

Deciduous teeth:
1D - Central incisor
2D - Lateral incisor
3D - Cuspid (canine)
4D - First molar (bicuspid)
5D - Second molar

Permanent teeth:
1P - Central incisor
2P - Lateral incisor
3P - Cuspid (canine)
4P - First premolar (bicuspid)
5P - Second premolar
6P - First molar
7P - Second molar

(a) Mandible of a six year old child showing erupted deciduous teeth and unerupted permanent teeth

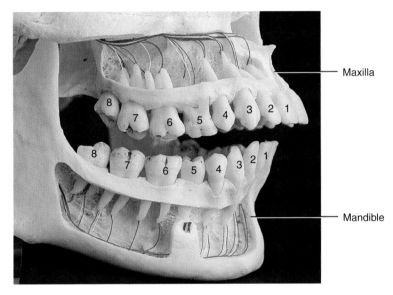

Right lateral view

1 - Central incisor
2 - Lateral incisor
3 - Cuspid (canine)
4 - First premolar (bicuspid)
5 - Second premolar
6 - First molar
7 - Second molar
8 - Third molar (wisdom tooth)

(b) Mandible (and maxilla) showing blood and nerve supply to permanent teeth

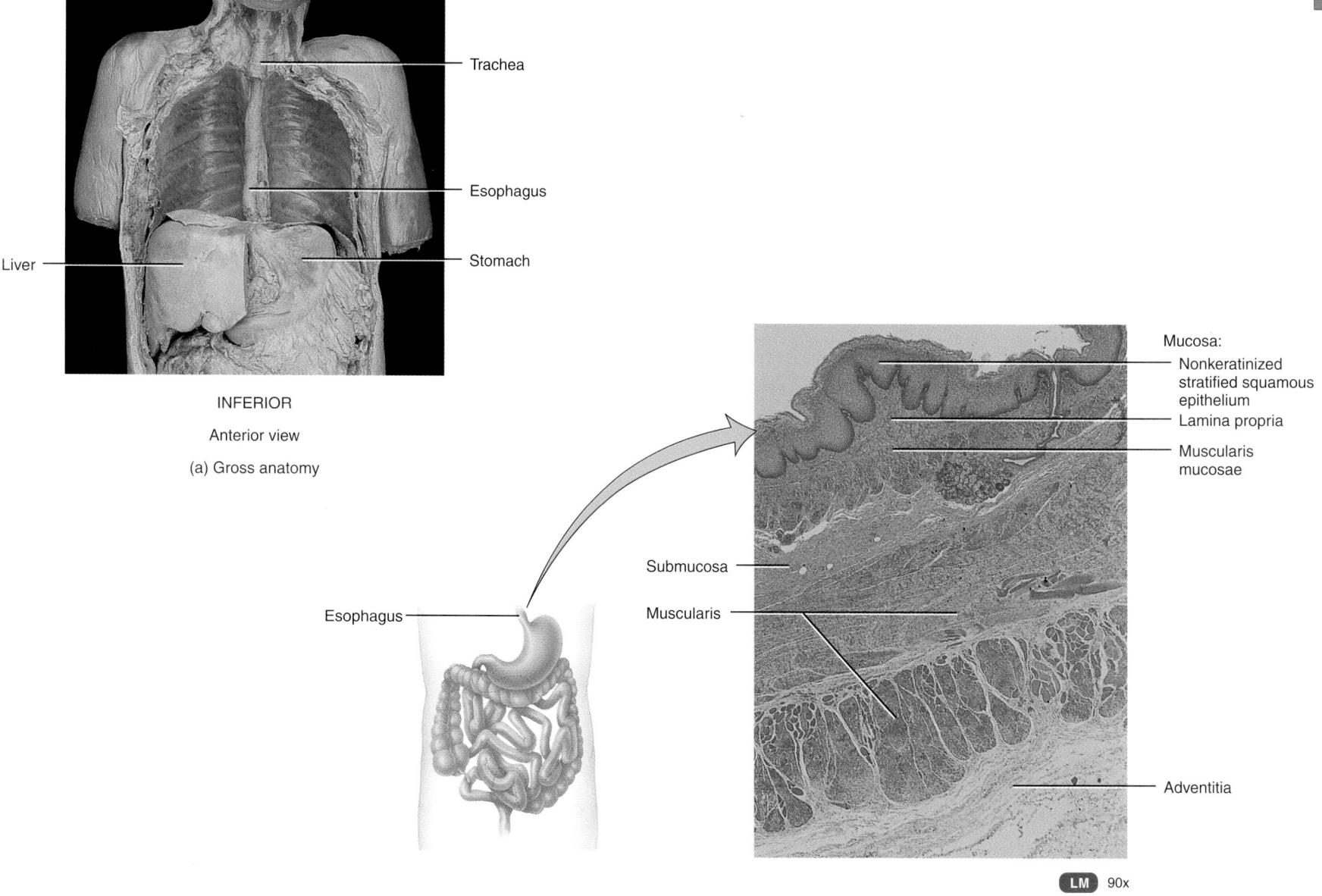

SUPERIOR

Trachea

Esophagus

Liver

Stomach

INFERIOR

Anterior view

(a) Gross anatomy

Esophagus

Mucosa:

Nonkeratinized stratified squamous epithelium

Lamina propria

Muscularis mucosae

Submucosa

Muscularis

Adventitia

LM 90x

(b) Histology of a portion of the wall of the esophagus

FIGURE 12.8 | *Esophagus*

SUPERIOR

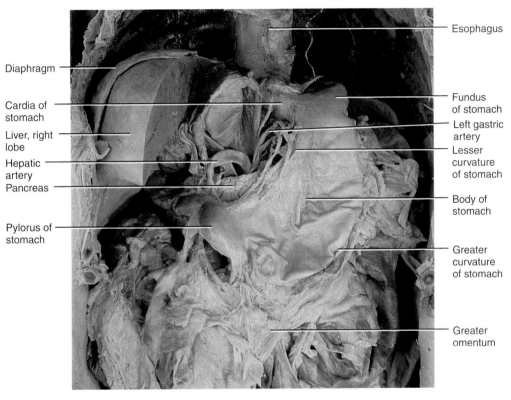

Esophagus

Diaphragm

Cardia of
stomach

Fundus
of stomach

Left gastric
artery

Liver, right
lobe

Lesser
curvature
of stomach

Hepatic
artery

Pancreas

Body of
stomach

Pylorus of
stomach

Greater
curvature
of stomach

Greater
omentum

INFERIOR

(a) Anterior view of external anatomy

FIGURE 12.9 | *Stomach*

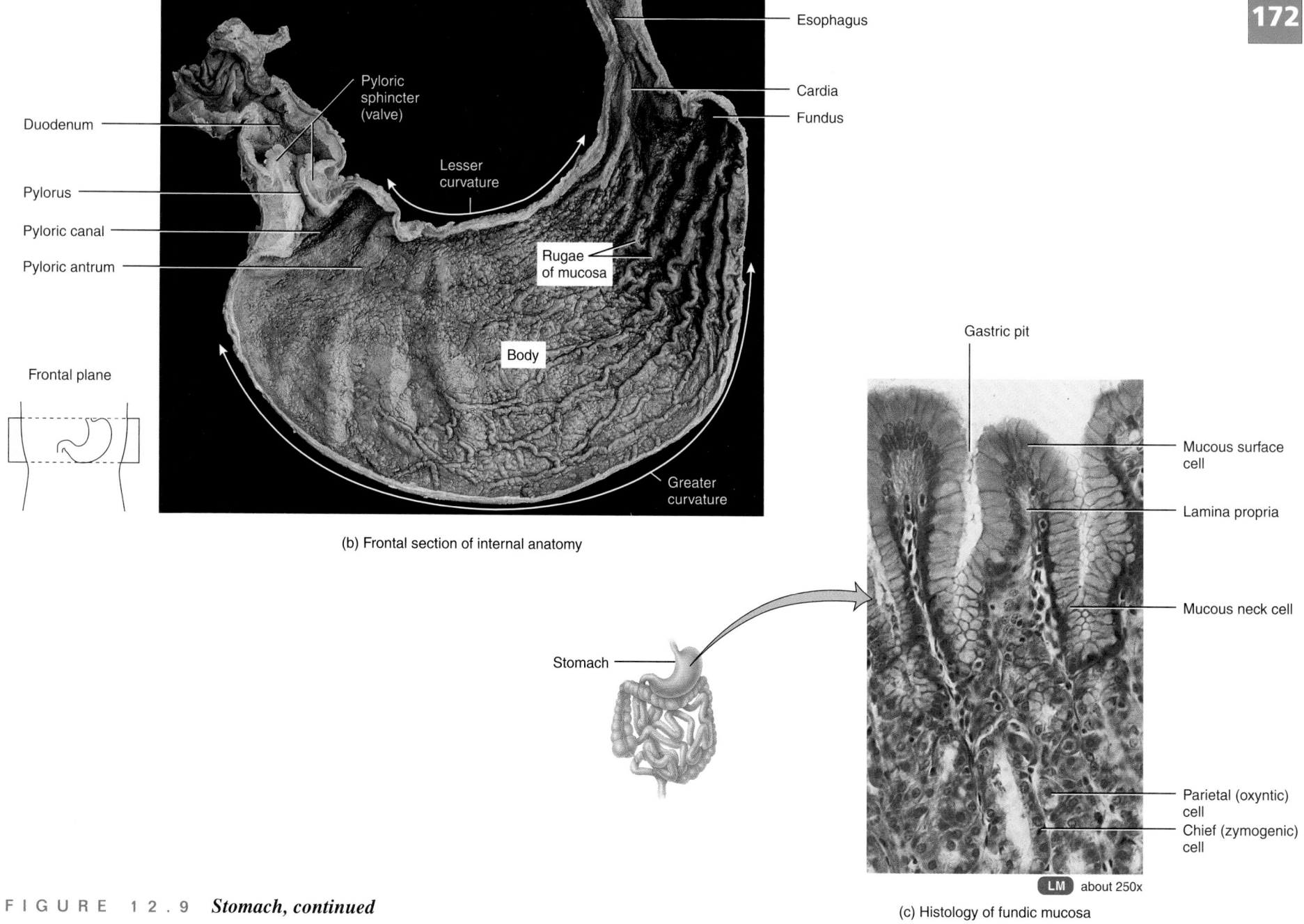

Esophagus

Cardia

Fundus

Pyloric sphincter (valve)

Duodenum

Pylorus

Pyloric canal

Pyloric antrum

Lesser curvature

Rugae of mucosa

Body

Greater curvature

Frontal plane

(b) Frontal section of internal anatomy

Stomach

Gastric pit

Mucous surface cell

Lamina propria

Mucous neck cell

Parietal (oxyntic) cell

Chief (zymogenic) cell

LM about 250x

(c) Histology of fundic mucosa

FIGURE  12.9  *Stomach, continued*

FIGURE 12.10 | *Pancreas*

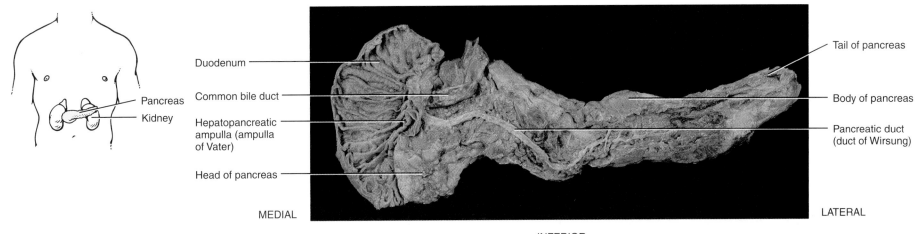

SUPERIOR

Duodenum

Common bile duct

Hepatopancreatic
ampulla (ampulla
of Vater)

Head of pancreas

Pancreas

Kidney

Tail of pancreas

Body of pancreas

Pancreatic duct
(duct of Wirsung)

MEDIAL

LATERAL

INFERIOR

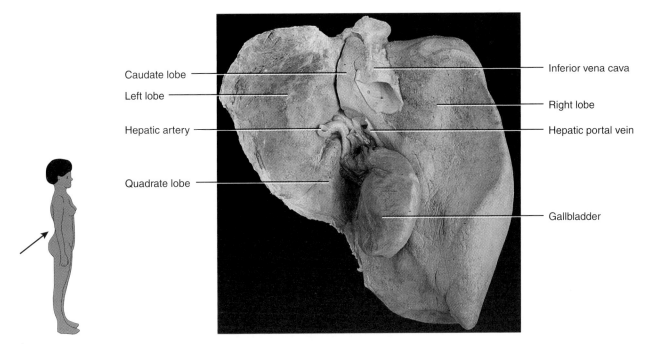

Caudate lobe

Left lobe

Hepatic artery

Quadrate lobe

Inferior vena cava

Right lobe

Hepatic portal vein

Gallbladder

FIGURE 12.11 | *Liver and gallbladder*

(a) Posteroinferior view of gross anatomy

**173**

Transverse
plane
through
abdomen

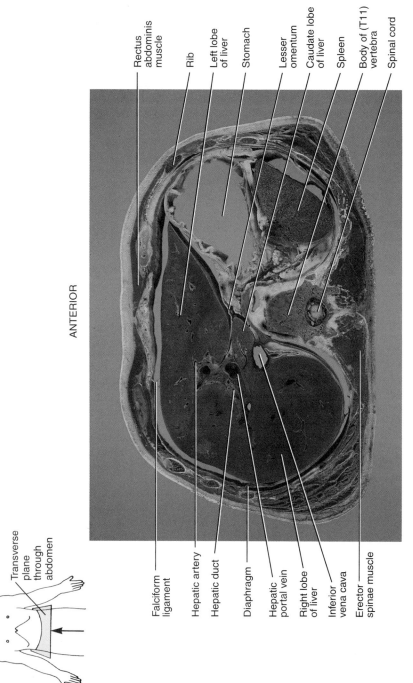

ANTERIOR

Rectus
abdominis
muscle

Rib

Left lobe
of liver

Stomach

Lesser
omentum

Caudate lobe
of liver

Spleen

Body of (T11)
vertebra

Spinal cord

Falciform
ligament

Hepatic artery

Hepatic duct

Diaphragm

Hepatic
portal vein

Right lobe
of liver

Inferior
vena cava

Erector
spinae muscle

POSTERIOR

(b) Inferior view of a transverse section of the abdomen

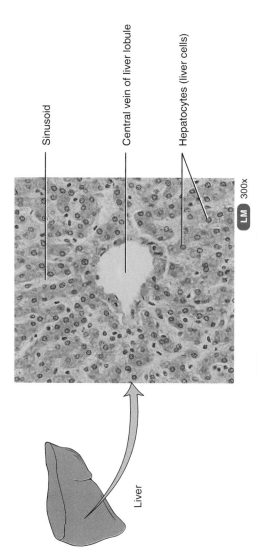

Sinusoid

Central vein of liver lobule

Hepatocytes (liver cells)

LM 300x

Liver

(c) Histology of a portion of a liver lobule

FIGURE 12.11 *Liver and gallbladder,
continued*

SUPERIOR

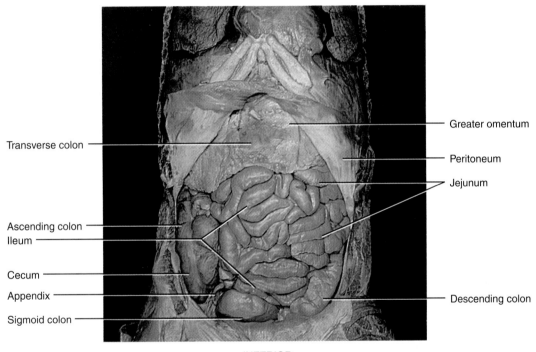

Transverse colon

Greater omentum

Peritoneum

Jejunum

Ascending colon

Ileum

Cecum

Appendix

Sigmoid colon

Descending colon

INFERIOR

(a) Anterior view of gross anatomy

Circular folds
(plicae circulares)

(b) Jejunum cut open to expose the circular folds

FIGURE 12.12 | *Small intestine*

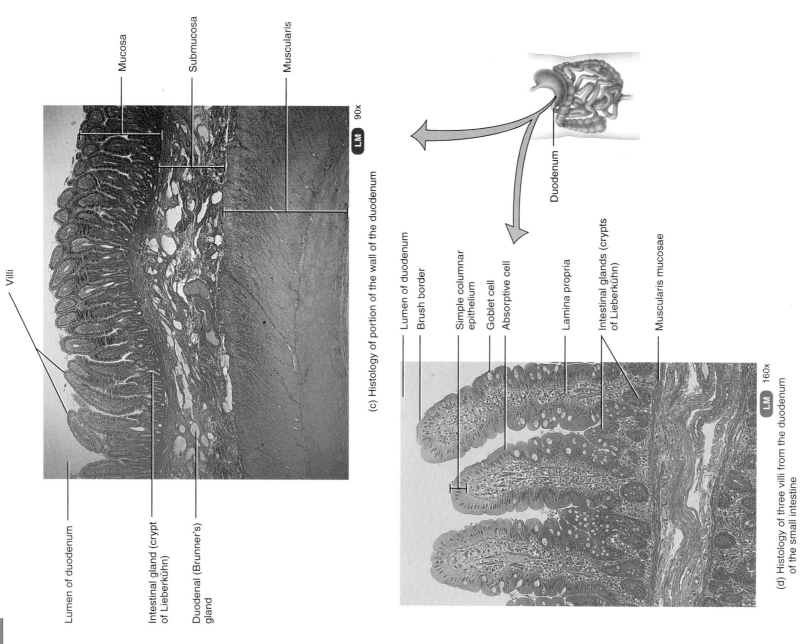

Mucosa

Submucosa

Muscularis

**LM** 90x

(c) Histology of portion of the wall of the duodenum

Villi

Lumen of duodenum

Intestinal gland (crypt of Lieberkühn)

Duodenal (Brunner's) gland

Duodenum

Lumen of duodenum

Brush border

Simple columnar epithelium

Goblet cell

Absorptive cell

Lamina propria

Intestinal glands (crypts of Lieberkühn)

Muscularis mucosae

**LM** 160x

(d) Histology of three villi from the duodenum of the small intestine

F I G U R E   1 2 . 1 2   |   *Small intestine, continued*

SUPERIOR

Liver ——

Ascending
colon ——

Cecum ——
Ileum ——

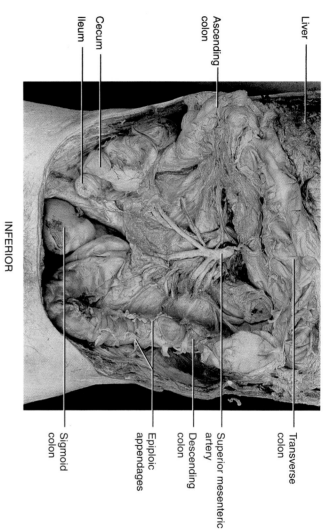

INFERIOR

(a) Anterior view of gross anatomy

Transverse
colon

Superior mesenteric
artery

Descending
colon

Epiploic
appendages

Sigmoid
colon

Goblet cell ——

**LM** 90x

(b) Histology of a portion of the wall
of the large intestine

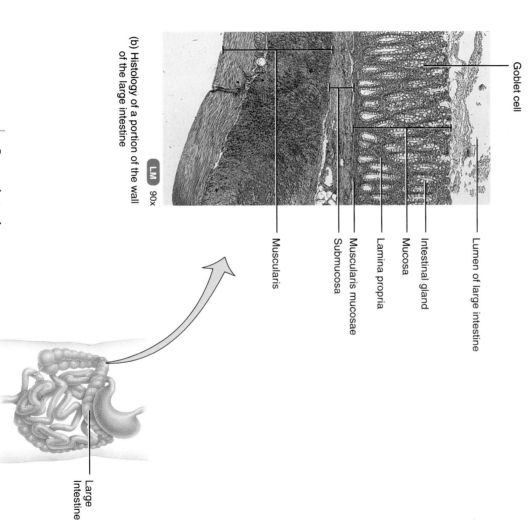

Muscularis

Submucosa

Muscularis mucosae

Lamina propria

Mucosa

Intestinal gland

Lumen of large intestine

Large
Intestine

FIGURE 12.13 | *Large intestine*

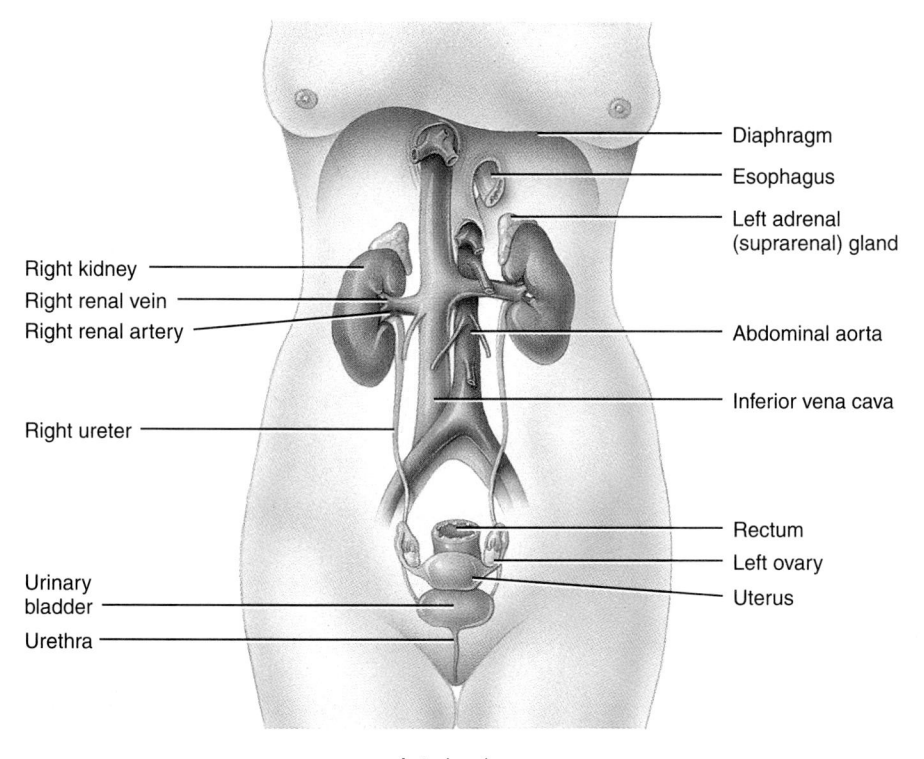

Right kidney

Right renal vein

Right renal artery

Right ureter

Urinary
bladder

Urethra

Diaphragm

Esophagus

Left adrenal
(suprarenal) gland

Abdominal aorta

Inferior vena cava

Rectum

Left ovary

Uterus

Anterior view

FIGURE 13.1 | *Urinary system*

SUPERIOR

Diaphragm

Right kidney
(internal view)

Inferior vena cava

Right renal artery

Right renal vein

Right ureter

Right testicular
vein

Right iliacus muscle

Right common
iliac artery

Right internal iliac vein

Right internal iliac
artery

Right external iliac
artery

Right ductus (vas)
deferens

Right external iliac vein

Left adrenal
(suprarenal)
gland

Left renal vein

Left kidney
(external view)

Abdominal aorta

Left psoas major
muscle

Left common iliac
vein

Left ureter

Urinary bladder

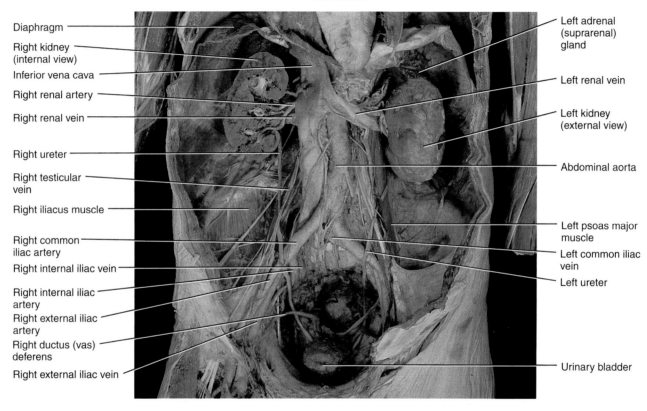

INFERIOR

Anterior view

FIGURE 13.2 | *Urinary organs*

SUPERIOR

Left renal vein

Left kidney (external view)

Abdominal aorta

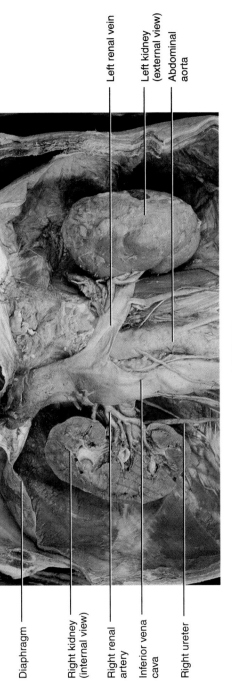

INFERIOR

Anterior view

Diaphragm

Right kidney (internal view)

Right renal artery

Inferior vena cava

Right ureter

FIGURE 13.3 | *Kidneys*

Adrenal (suprarenal) gland

Inferior vena cava

Suprarenal arteries

Renal artery

Renal vein

Kidney

Ureter

SUPERIOR

MEDIAL

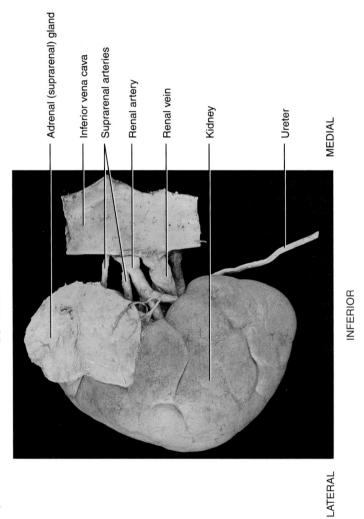

INFERIOR

Anterior view

LATERAL

FIGURE 13.4 | *Right kidney, external aspect*

F I G U R E 1 3 . 5 | *Right kidney, internal aspect. A transverse section of the kidneys is illustrated in Fig. 12.4.*

181

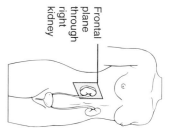

Frontal plane through right kidney

SUPERIOR

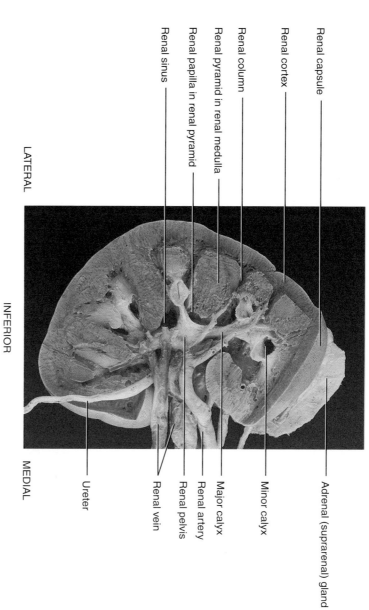

LATERAL

INFERIOR

MEDIAL

(a) Human kidney

Anterior view

Renal capsule

Renal cortex

Renal column

Renal pyramid in renal medulla

Renal papilla in renal pyramid

Renal sinus

Renal capsule

Minor calyx

Major calyx

Renal artery

Renal pelvis

Renal vein

Ureter

Adrenal (suprarenal) gland

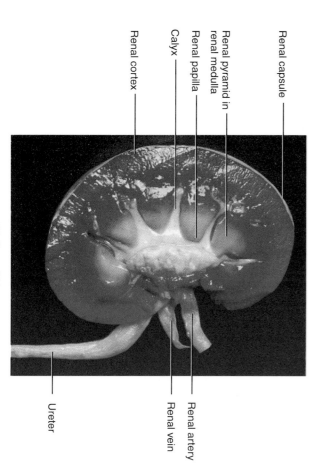

(b) Sheep kidney

Renal capsule

Renal pyramid in renal medulla

Renal papilla

Calyx

Renal cortex

Renal artery

Renal vein

Ureter

SUPERIOR

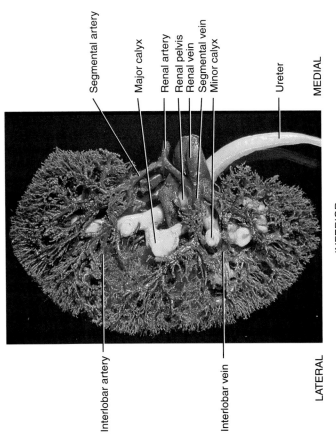

Segmental artery

Major calyx

Renal artery
Renal pelvis
Renal vein
Segmental vein
Minor calyx

Ureter

MEDIAL

Interlobar artery

Interlobar vein

LATERAL

INFERIOR

Anterior view

*Right kidney, cast of blood supply. Blood vessels are red and blue; urine-draining structures are yellow.*

FIGURE 13.6

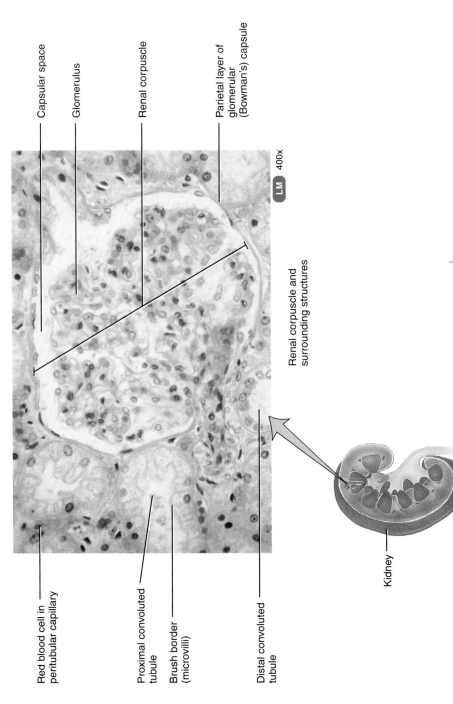

Capsular space

Glomerulus

Renal corpuscle

Parietal layer of glomerular (Bowman's) capsule

LM 400x

Renal corpuscle and surrounding structures

Red blood cell in peritubular capillary

Proximal convoluted tubule

Brush border (microvilli)

Distal convoluted tubule

Kidney

FIGURE 13.7 | *Histology of the kidney*

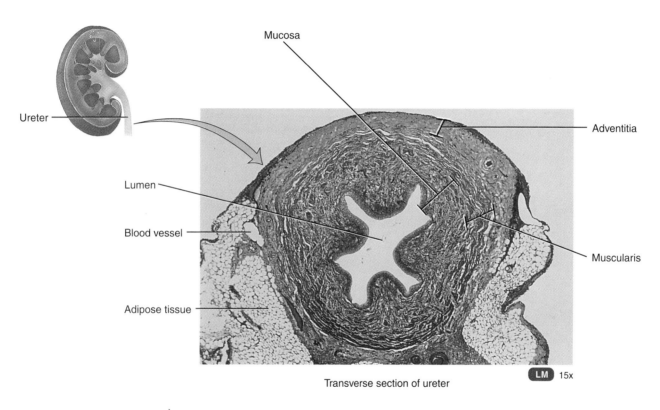

Mucosa

Adventitia

Ureter

Lumen

Blood vessel

Muscularis

Adipose tissue

LM  15x

Transverse section of ureter

FIGURE  13.8  |  *Histology of the ureter*

FIGURE 13.9 | *Urinary bladder*

Midsagittal plane

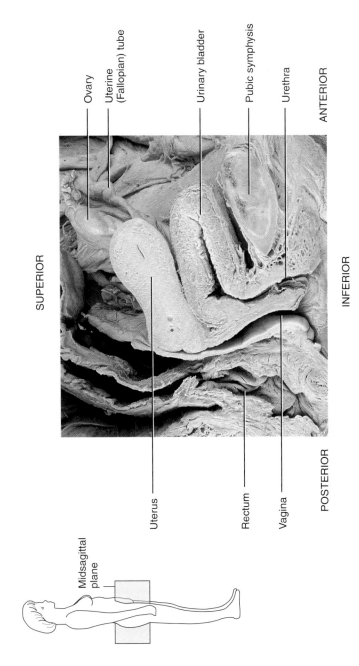

SUPERIOR

Ovary

Uterine (Fallopian) tube

Urinary bladder

Pubic symphysis

Urethra

ANTERIOR

POSTERIOR

Uterus

Rectum

Vagina

INFERIOR

(a) Midsagittal section of gross anatomy

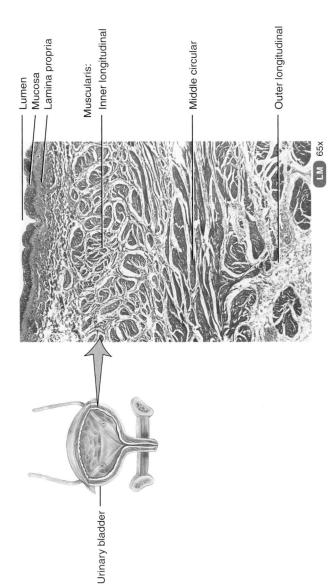

Lumen
Mucosa
Lamina propria

Muscularis:
Inner longitudinal

Middle circular

Outer longitudinal

LM 65x

Urinary bladder

(b) Histology of a portion of the wall of the urinary bladder

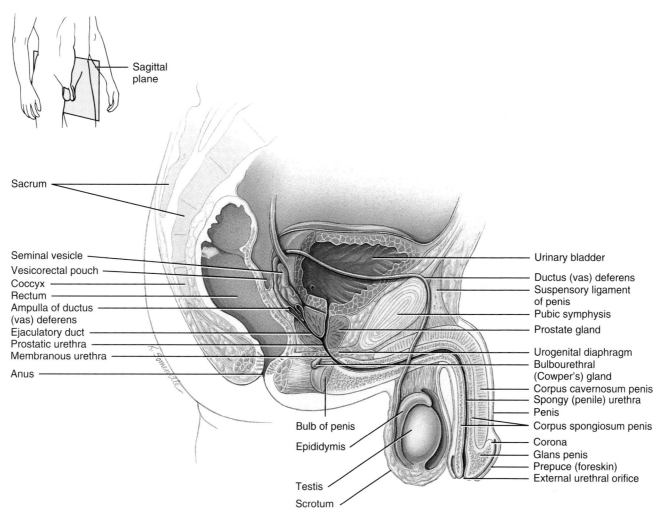

Sagittal plane

Sacrum

Seminal vesicle
Vesicorectal pouch
Coccyx
Rectum
Ampulla of ductus (vas) deferens
Ejaculatory duct
Prostatic urethra
Membranous urethra

Anus

Bulb of penis

Epididymis

Testis

Scrotum

Urinary bladder

Ductus (vas) deferens
Suspensory ligament of penis

Pubic symphysis

Prostate gland

Urogenital diaphragm
Bulbourethral (Cowper's) gland
Corpus cavernosum penis
Spongy (penile) urethra
Penis
Corpus spongiosum penis
Corona
Glans penis
Prepuce (foreskin)
External urethral orifice

(a) Sagittal section of male reproductive system

FIGURE 14.1 | *Reproductive systems*

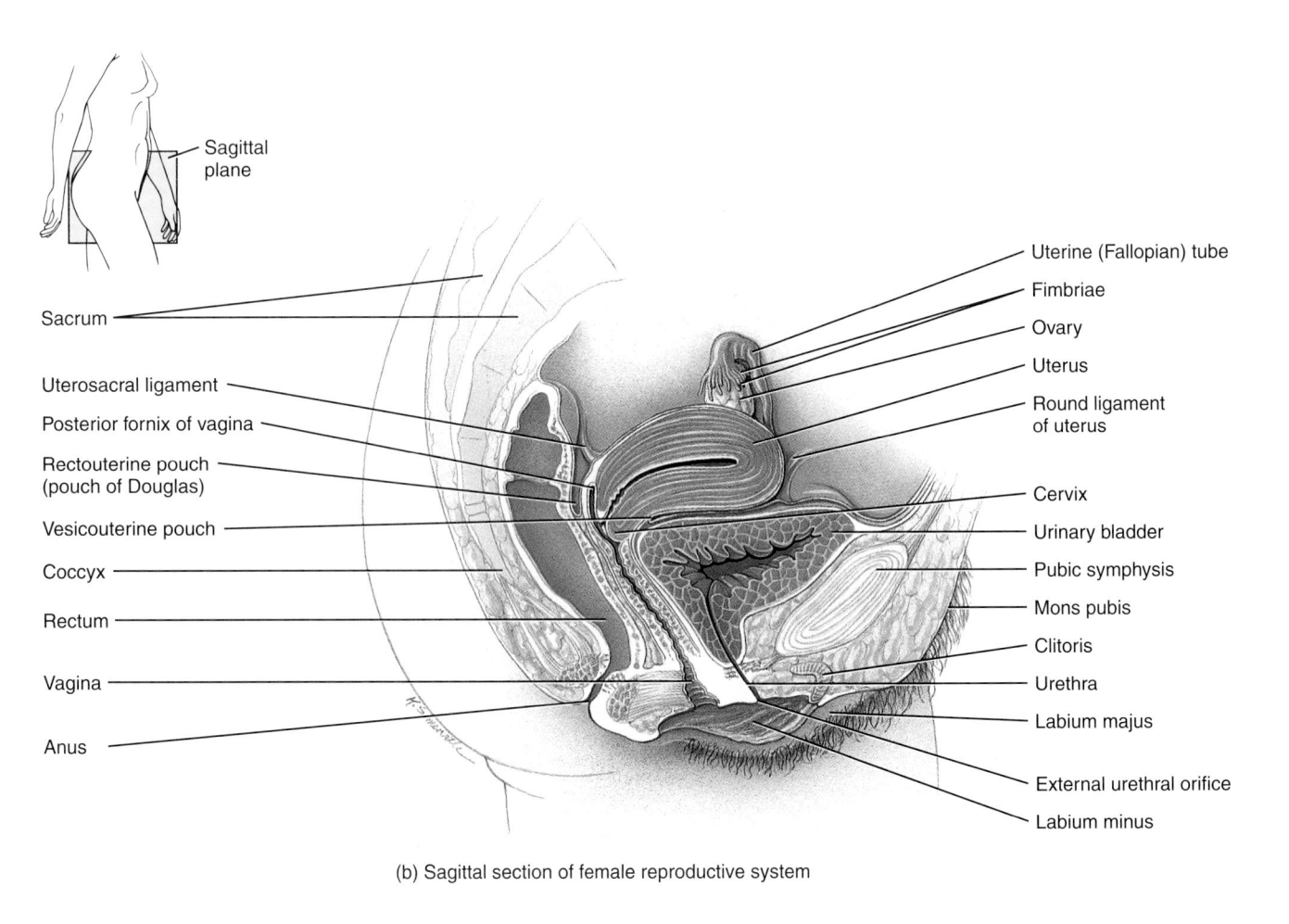

Sagittal plane

Sacrum

Uterosacral ligament

Posterior fornix of vagina

Rectouterine pouch
(pouch of Douglas)

Vesicouterine pouch

Coccyx

Rectum

Vagina

Anus

Uterine (Fallopian) tube

Fimbriae

Ovary

Uterus

Round ligament
of uterus

Cervix

Urinary bladder

Pubic symphysis

Mons pubis

Clitoris

Urethra

Labium majus

External urethral orifice

Labium minus

(b) Sagittal section of female reproductive system

FIGURE 14.1    *Reproductive systems,*
*continued*

SUPERIOR

POSTERIOR

ANTERIOR

Ureter

Urinary bladder
(opened)

Ductus (vas)
deferens

Ureter

Seminal vesicle
(sectioned)

Ampulla of ductus
(vas) deferens

Prostatic urethra

Ejaculatory duct

Prostate gland

Crus of penis
covered by
ischiocavernosus
muscle

Pubic symphysis

Corpus
cavernosum penis

Corpus
spongiosum penis

Spongy (penile)
urethra

Corona

Glans penis

Bulbospongiosus
muscle

Bulb of penis

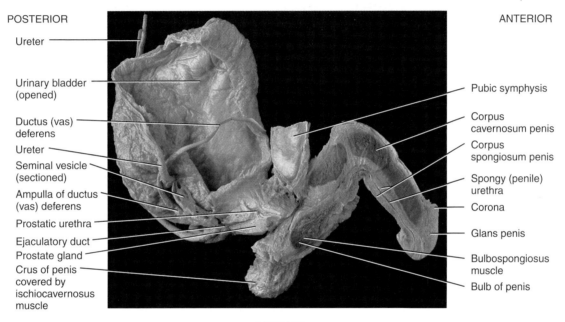

INFERIOR

Sagittal dissection

FIGURE 14.2 | *Male reproductive organs*

187

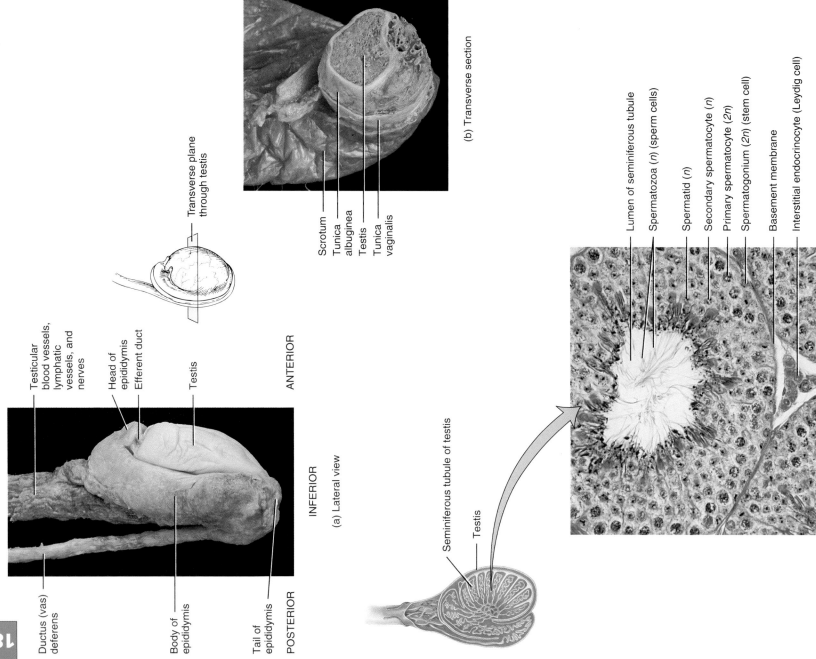

**188**

SUPERIOR

Testicular blood vessels, lymphatic vessels, and nerves

Head of epididymis

Efferent duct

Testis

Ductus (vas) deferens

Body of epididymis

Tail of epididymis

POSTERIOR

ANTERIOR

INFERIOR

(a) Lateral view

Transverse plane through testis

Scrotum
Tunica albuginea
Testis
Tunica vaginalis

(b) Transverse section

Seminiferous tubule of testis

Testis

Lumen of seminiferous tubule

Spermatozoa (*n*) (sperm cells)

Spermatid (*n*)

Secondary spermatocyte (*n*)

Primary spermatocyte (*2n*)

Spermatogonium (*2n*) (stem cell)

Basement membrane

Interstitial endocrinocyte (Leydig cell)

LM  310x

(c) Histology of several seminiferous tubules

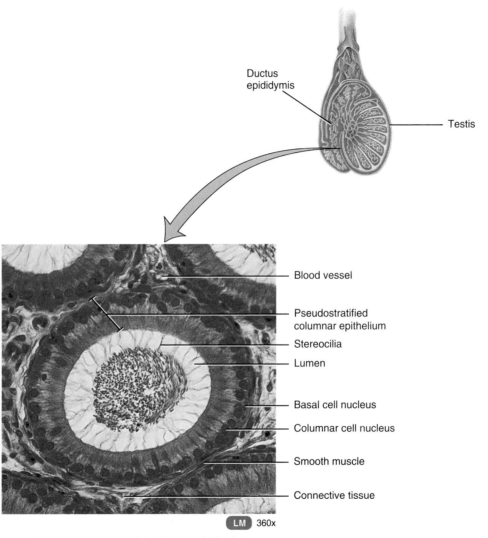

Ductus epididymis

Testis

Blood vessel

Pseudostratified columnar epithelium

Stereocilia

Lumen

Basal cell nucleus

Columnar cell nucleus

Smooth muscle

Connective tissue

LM 360x

Transverse section of the ductus epididymis

FIGURE 14.4 | *Histology of the ductus epididymis*

FIGURE 14.5 | *Histology of the ductus (vas) deferens*

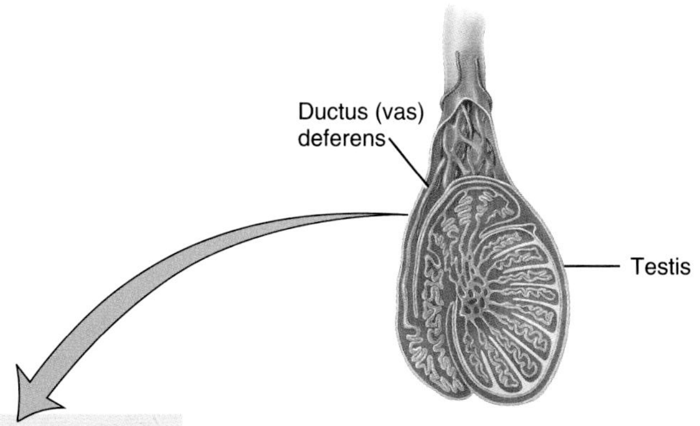

190

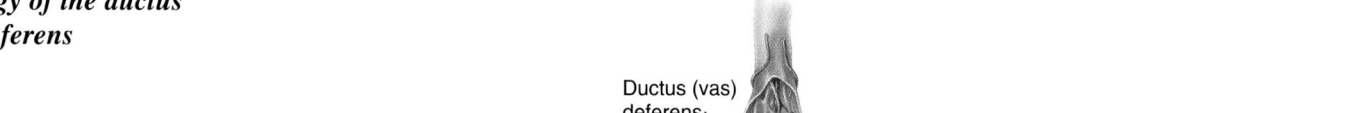

Ductus (vas) deferens

Testis

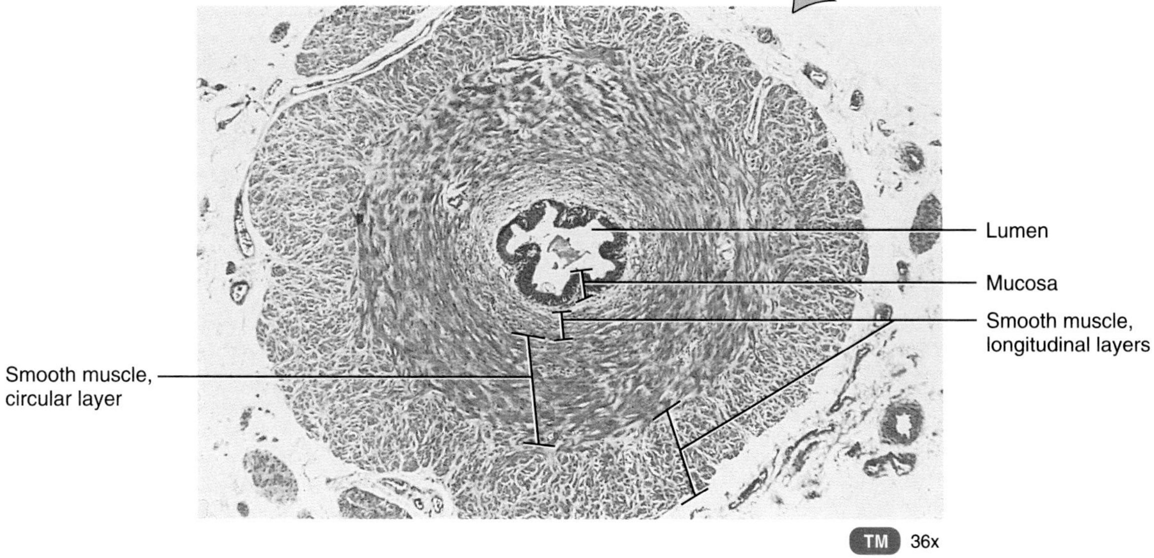

Lumen

Mucosa

Smooth muscle, longitudinal layers

Smooth muscle, circular layer

TM 36x

(a) Transverse section of ductus (vas) deferens

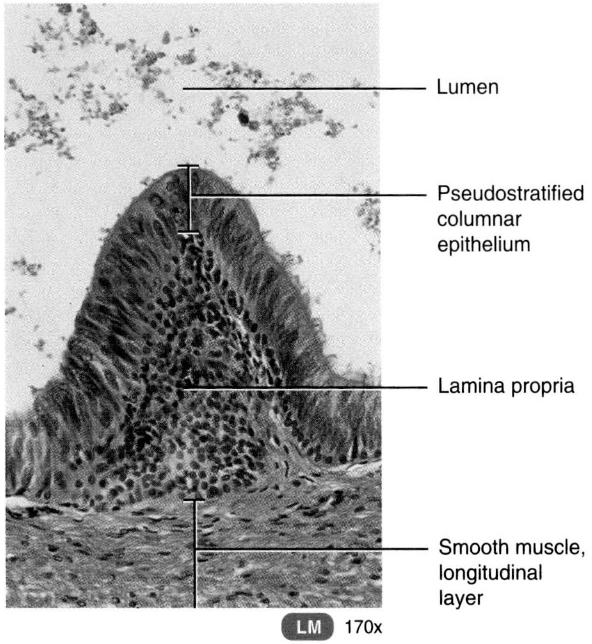

Lumen

Pseudostratified columnar epithelium

Lamina propria

Smooth muscle, longitudinal layer

LM 170x

(b) Transverse section of mucosa of ductus (vas) deferens

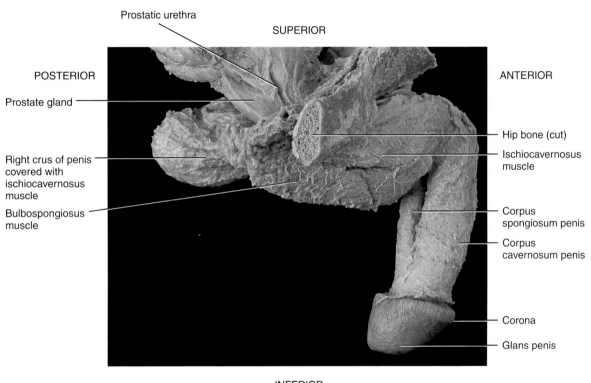

Prostatic urethra

SUPERIOR

POSTERIOR

ANTERIOR

Prostate gland

Hip bone (cut)

Right crus of penis
covered with
ischiocavernosus
muscle

Ischiocavernosus
muscle

Bulbospongiosus
muscle

Corpus
spongiosum penis

Corpus
cavernosum penis

Corona

Glans penis

INFERIOR

Right lateral view

FIGURE 14.6 | *Penis*

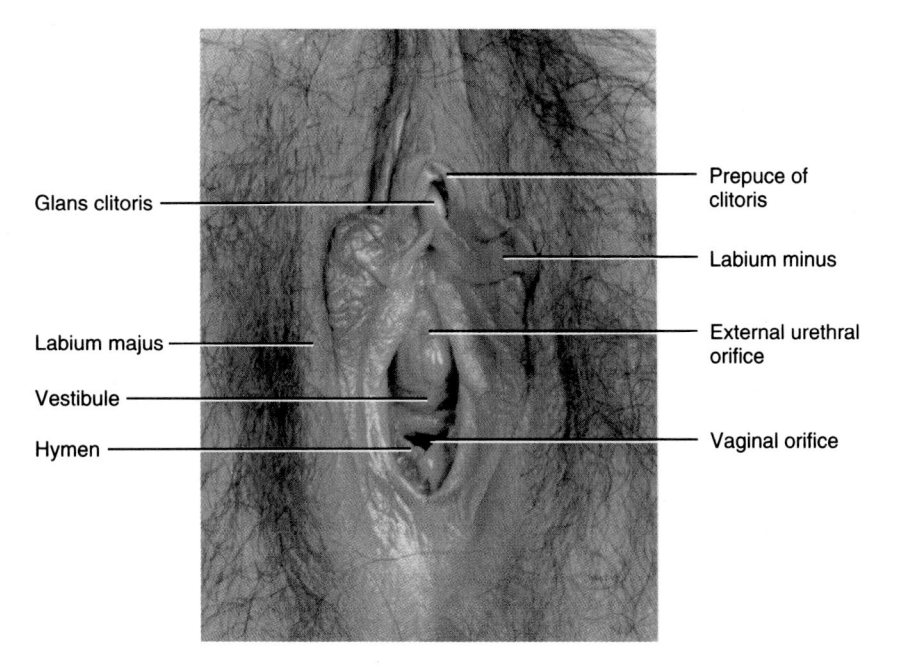

Glans clitoris

Labium majus

Vestibule

Hymen

Prepuce of clitoris

Labium minus

External urethral orifice

Vaginal orifice

Inferior view

FIGURE 14.7 | *Female external genitals (vulva)*

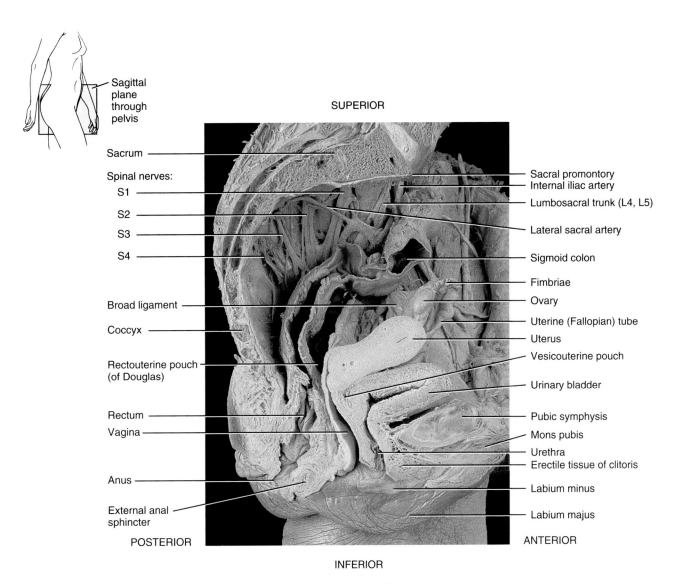

Sagittal plane through pelvis

Sacrum

Spinal nerves:
 S1
 S2
 S3
 S4

Broad ligament

Coccyx

Rectouterine pouch (of Douglas)

Rectum

Vagina

Anus

External anal sphincter

Sacral promontory
Internal iliac artery

Lumbosacral trunk (L4, L5)

Lateral sacral artery

Sigmoid colon

Fimbriae

Ovary

Uterine (Fallopian) tube

Uterus

Vesicouterine pouch

Urinary bladder

Pubic symphysis

Mons pubis

Urethra
Erectile tissue of clitoris

Labium minus

Labium majus

POSTERIOR

ANTERIOR

INFERIOR

(a) Sagittal section

FIGURE 14.8 | *Female reproductive organs*

FIGURE 14.9 | **Uterus and associated structures**

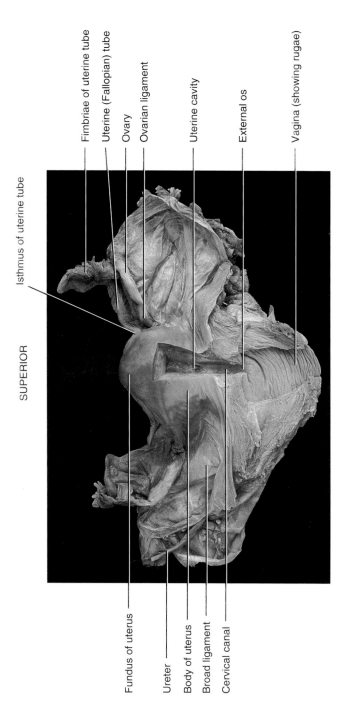

Fimbriae of uterine tube
Uterine (Fallopian) tube
Ovary
Ovarian ligament
Uterine cavity
External os
Vagina (showing rugae)

Isthmus of uterine tube

SUPERIOR

INFERIOR

Fundus of uterus
Ureter
Body of uterus
Broad ligament
Cervical canal

(a) Posterior view

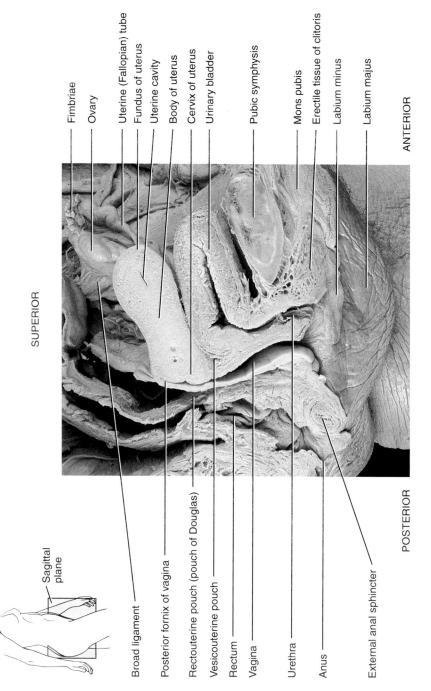

Fimbriae
Ovary
Uterine (Fallopian) tube
Fundus of uterus
Uterine cavity
Body of uterus
Cervix of uterus
Urinary bladder
Pubic symphysis
Mons pubis
Erectile tissue of clitoris
Labium minus
Labium majus

ANTERIOR

SUPERIOR

INFERIOR

POSTERIOR

Sagittal plane

Broad ligament
Posterior fornix of vagina
Rectouterine pouch (pouch of Douglas)
Vesicouterine pouch
Rectum
Vagina
Urethra
Anus
External anal sphincter

(b) Sagittal section

FIGURE 14.10 | *Histology of the ovary*

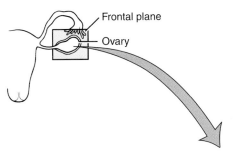
Frontal plane
Ovary

Primordial follicles  Germinal epithelium  Tunica albuginea  Ovarian cortex

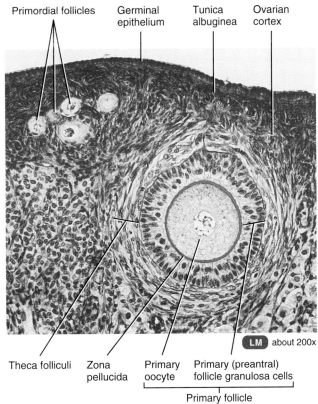

LM  about 200x

Theca folliculi   Zona pellucida   Primary oocyte   Primary (preantral) follicle granulosa cells

Primary follicle

(a) Ovarian cortex showing primordial follicles and a primary follicle

Antrum filled with follicular fluid   Corona radiata   Secondary (antral) follicle granulosa cells

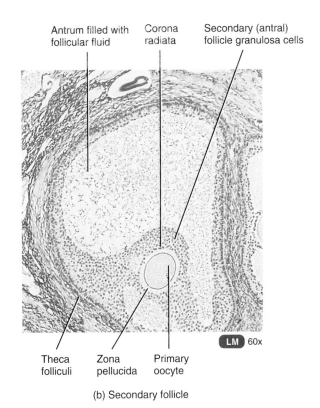

LM  60x

Theca folliculi   Zona pellucida   Primary oocyte

(b) Secondary follicle

195

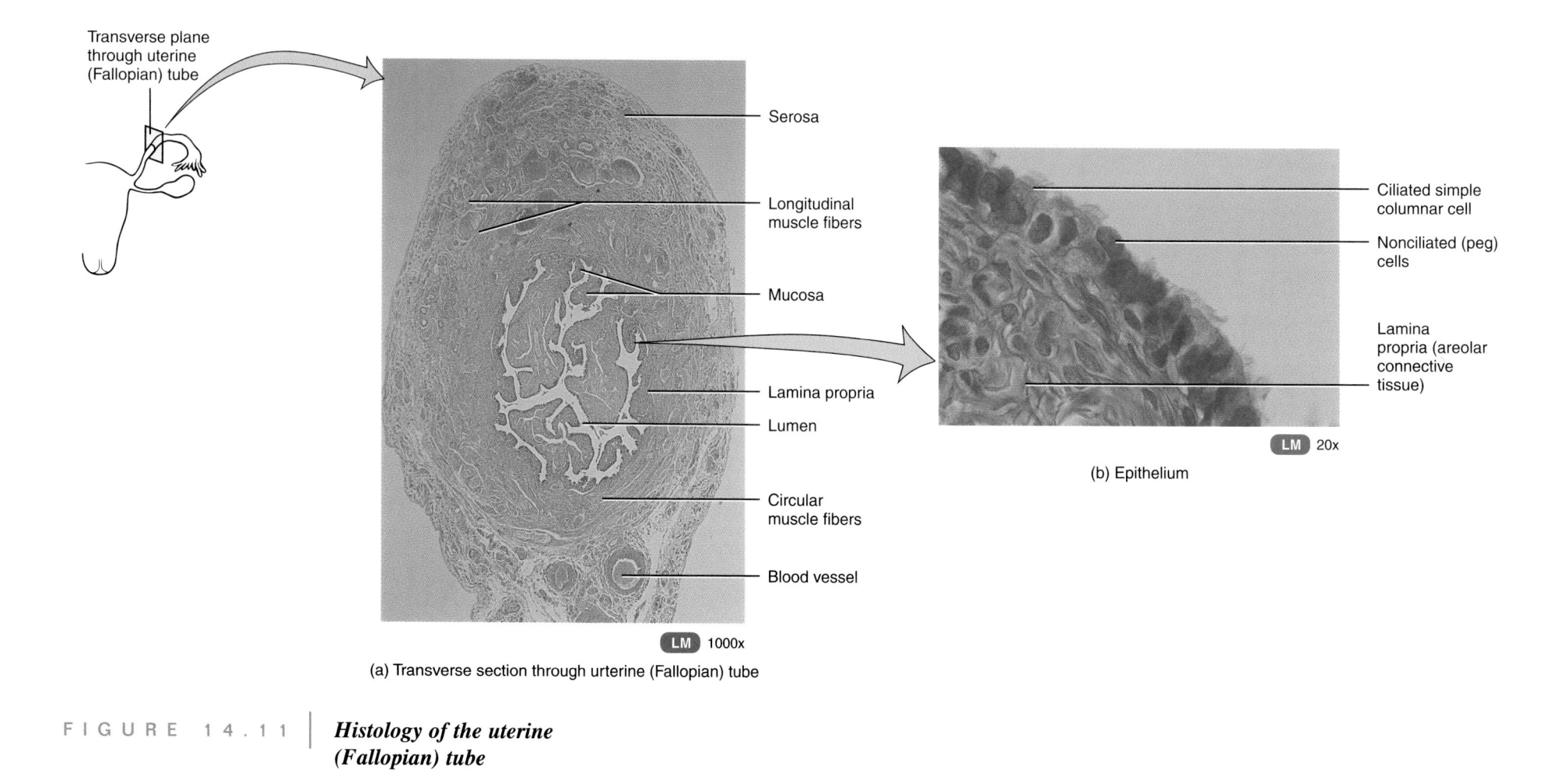

Transverse plane through uterine (Fallopian) tube

Serosa

Longitudinal muscle fibers

Mucosa

Lamina propria

Lumen

Circular muscle fibers

Blood vessel

**LM** 1000x

(a) Transverse section through urterine (Fallopian) tube

Ciliated simple columnar cell

Nonciliated (peg) cells

Lamina propria (areolar connective tissue)

**LM** 20x

(b) Epithelium

F I G U R E   1 4 . 1 1    *Histology of the uterine (Fallopian) tube*

FIGURE 14.12 *Histology of the uterus*

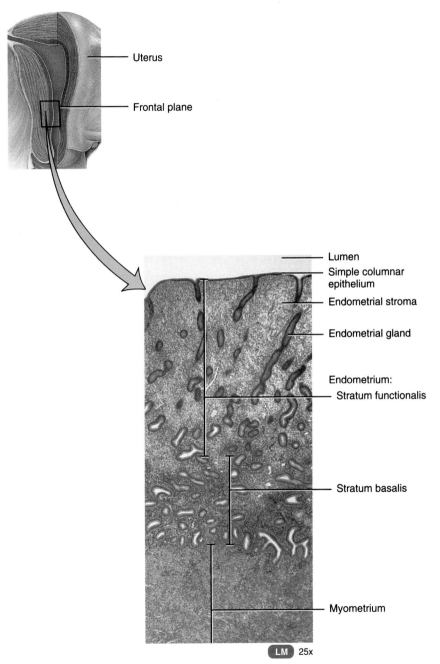

Uterus

Frontal plane

Lumen

Simple columnar epithelium

Endometrial stroma

Endometrial gland

Endometrium:
Stratum functionalis

Stratum basalis

Myometrium

LM 25x

Portion of endometrium and myometrium

197

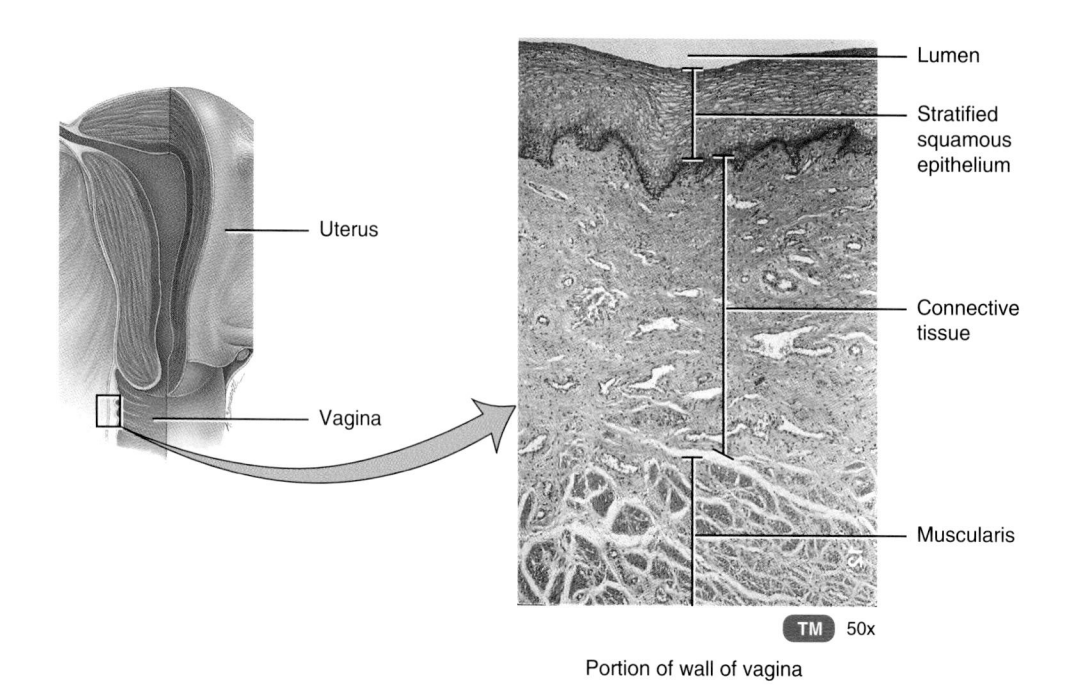

Uterus

Vagina

Lumen

Stratified
squamous
epithelium

Connective
tissue

Muscularis

TM 50x

Portion of wall of vagina

FIGURE 14.13 | *Histology of the vagina*

# FIGURE 14.14 | *Mammary glands*

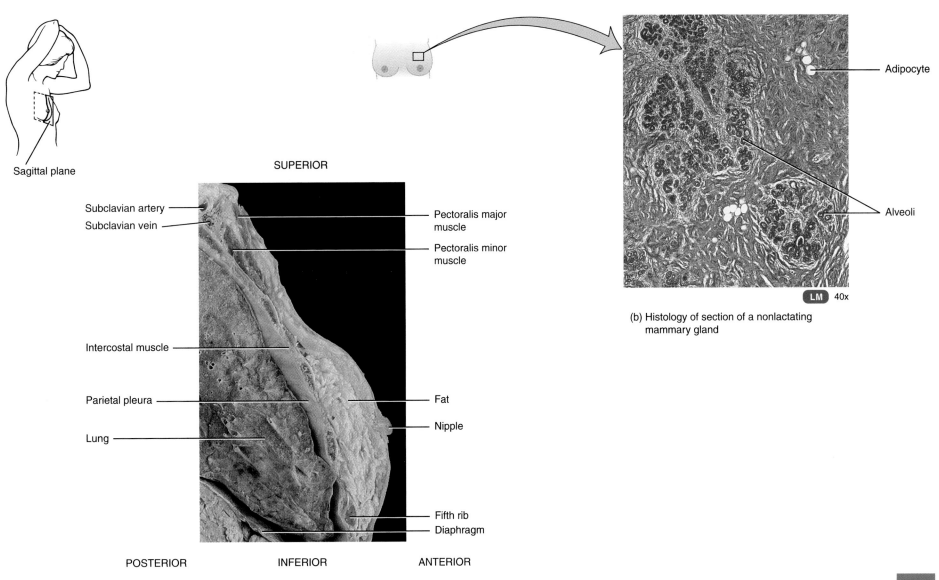

Sagittal plane

SUPERIOR

Subclavian artery

Subclavian vein

Pectoralis major
muscle

Pectoralis minor
muscle

Intercostal muscle

Parietal pleura — Fat

Lung — Nipple

Fifth rib
Diaphragm

POSTERIOR          INFERIOR          ANTERIOR

(a) Sagittal section of gross anatomy

Adipocyte

Alveoli

LM  40x

(b) Histology of section of a nonlactating
mammary gland

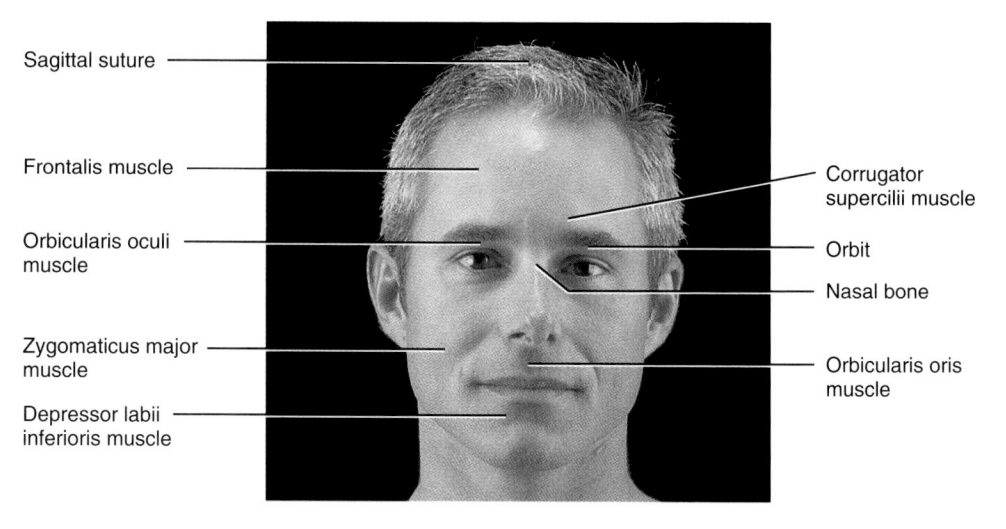

Sagittal suture

Frontalis muscle

Orbicularis oculi muscle

Zygomaticus major muscle

Depressor labii inferioris muscle

Corrugator supercilii muscle

Orbit

Nasal bone

Orbicularis oris muscle

(a) Anterior view of regions of the head

FIGURE 15.1 | *Surface anatomy of the head and neck. Refer to Fig. 9.3a for the surface anatomy of the eye, Fig. 9.4a for the surface anatomy of the ear, and 9.1a for the surface anatomy of the nose.*

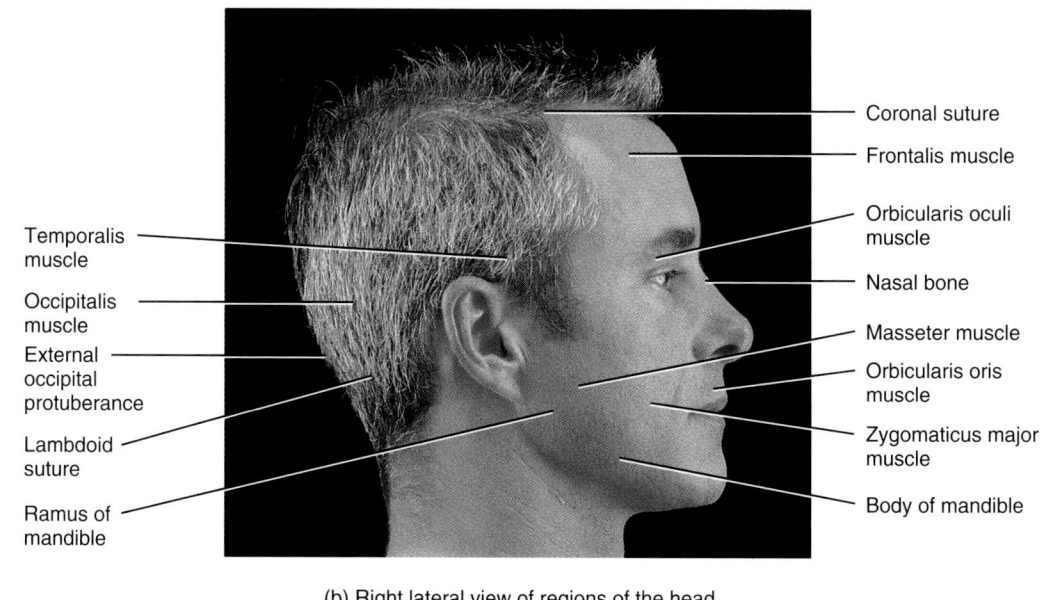

Coronal suture

Frontalis muscle

Orbicularis oculi muscle

Nasal bone

Masseter muscle

Orbicularis oris muscle

Zygomaticus major muscle

Body of mandible

Temporalis muscle

Occipitalis muscle

External occipital protuberance

Lambdoid suture

Ramus of mandible

(b) Right lateral view of regions of the head

FIGURE 15.1 **Surface anatomy of the head and neck, continued**

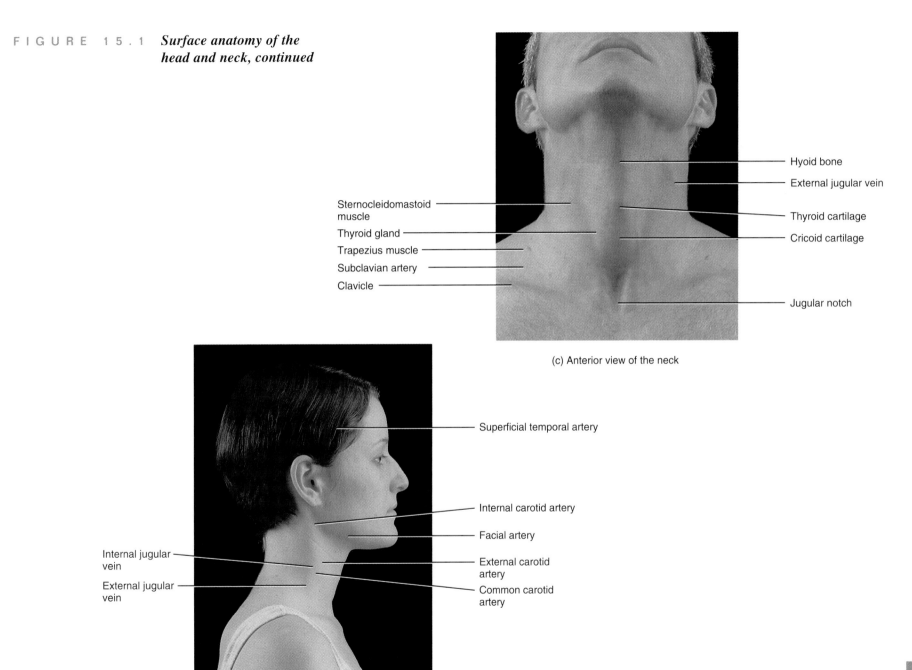

Hyoid bone

External jugular vein

Sternocleidomastoid muscle

Thyroid cartilage

Thyroid gland

Cricoid cartilage

Trapezius muscle

Subclavian artery

Clavicle

Jugular notch

(c) Anterior view of the neck

Superficial temporal artery

Internal carotid artery

Facial artery

Internal jugular vein

External carotid artery

External jugular vein

Common carotid artery

(d) Right lateral view of the neck

F I G U R E   1 5 . 2 | *Surface anatomy of the trunk*

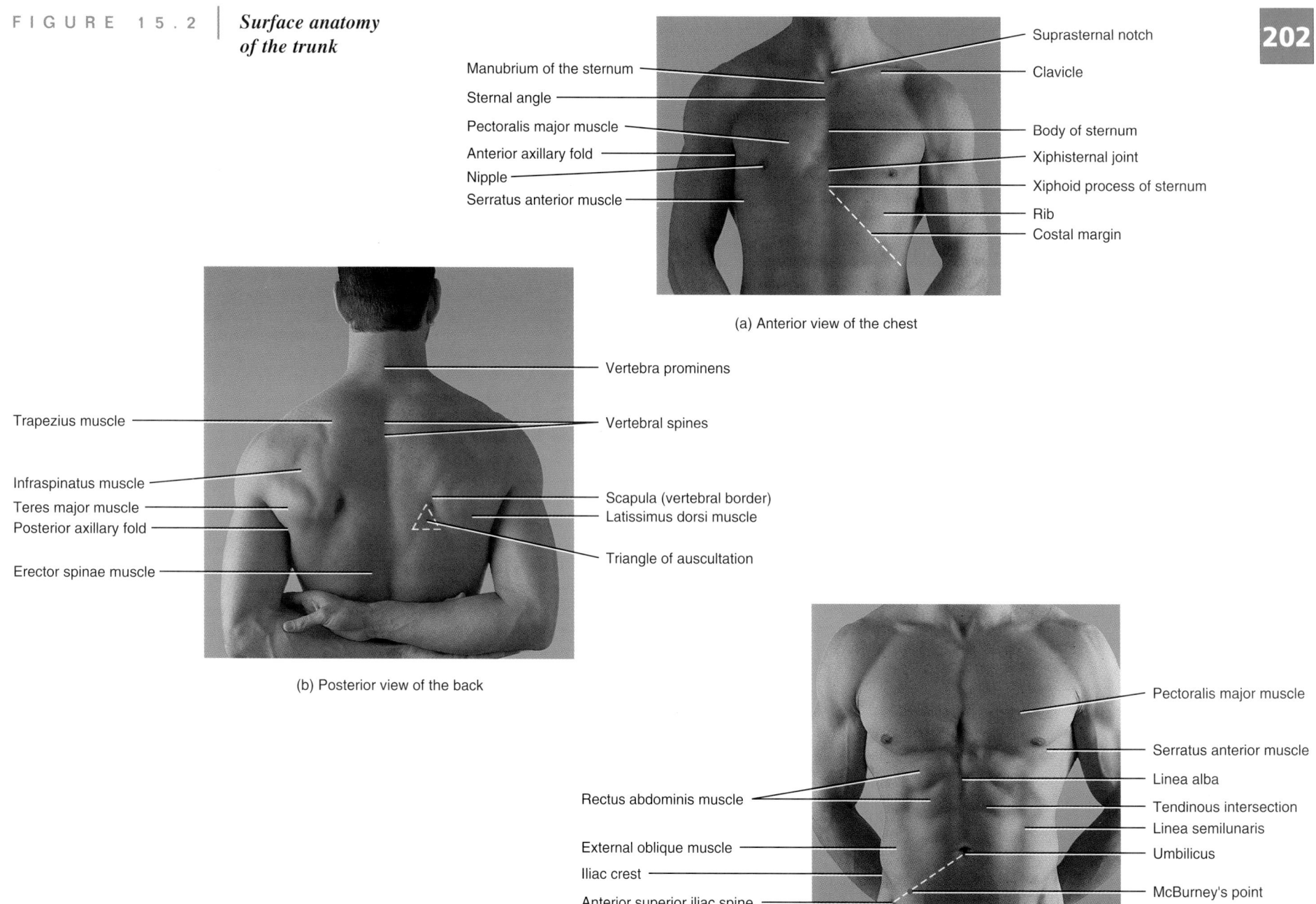

Manubrium of the sternum
Sternal angle
Pectoralis major muscle
Anterior axillary fold
Nipple
Serratus anterior muscle

Suprasternal notch
Clavicle
Body of sternum
Xiphisternal joint
Xiphoid process of sternum
Rib
Costal margin

(a) Anterior view of the chest

Trapezius muscle

Infraspinatus muscle
Teres major muscle
Posterior axillary fold

Erector spinae muscle

Vertebra prominens

Vertebral spines

Scapula (vertebral border)
Latissimus dorsi muscle

Triangle of auscultation

(b) Posterior view of the back

Rectus abdominis muscle

External oblique muscle
Iliac crest
Anterior superior iliac spine

Pectoralis major muscle

Serratus anterior muscle
Linea alba
Tendinous intersection
Linea semilunaris
Umbilicus

McBurney's point

(c) Anterior view of the abdomen

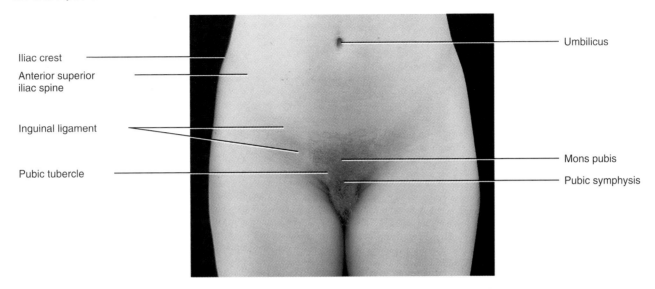

Iliac crest

Anterior superior
iliac spine

Inguinal ligament

Pubic tubercle

Umbilicus

Mons pubis

Pubic symphysis

(d) Anterior view of the pelvis

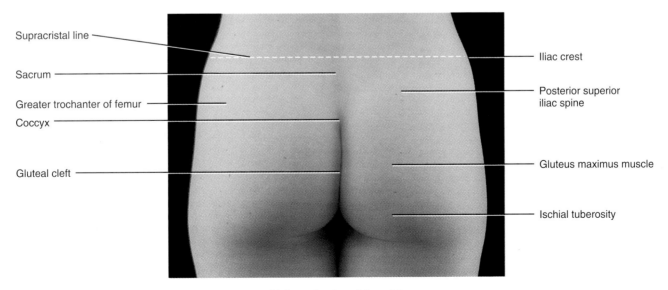

Supracristal line

Sacrum

Greater trochanter of femur

Coccyx

Gluteal cleft

Iliac crest

Posterior superior
iliac spine

Gluteus maximus muscle

Ischial tuberosity

(e) Posterior view of the pelvis

FIGURE 15.3 | *Surface anatomy of the upper limb*

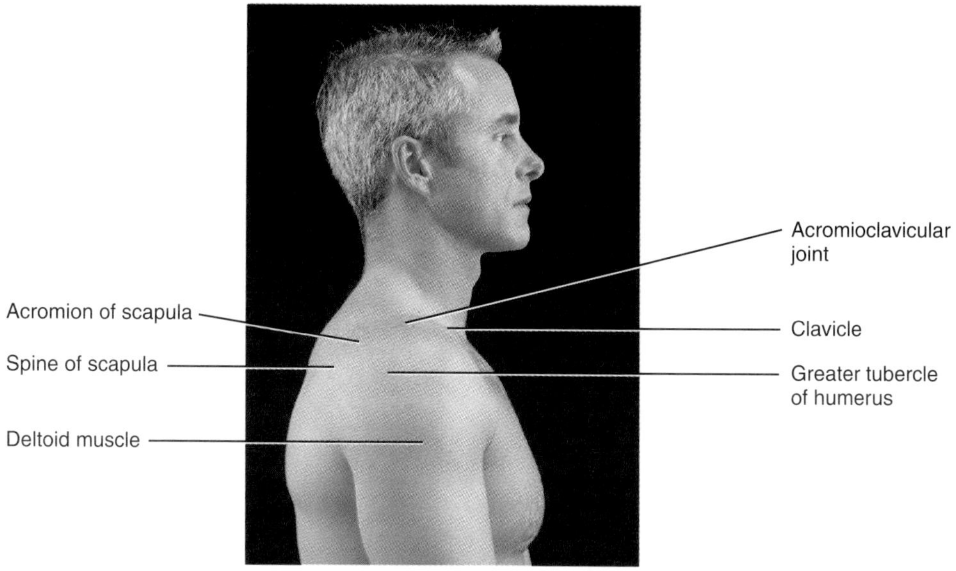

Acromioclavicular joint

Acromion of scapula

Spine of scapula

Deltoid muscle

Clavicle

Greater tubercle of humerus

(a) Right lateral view of the shoulder

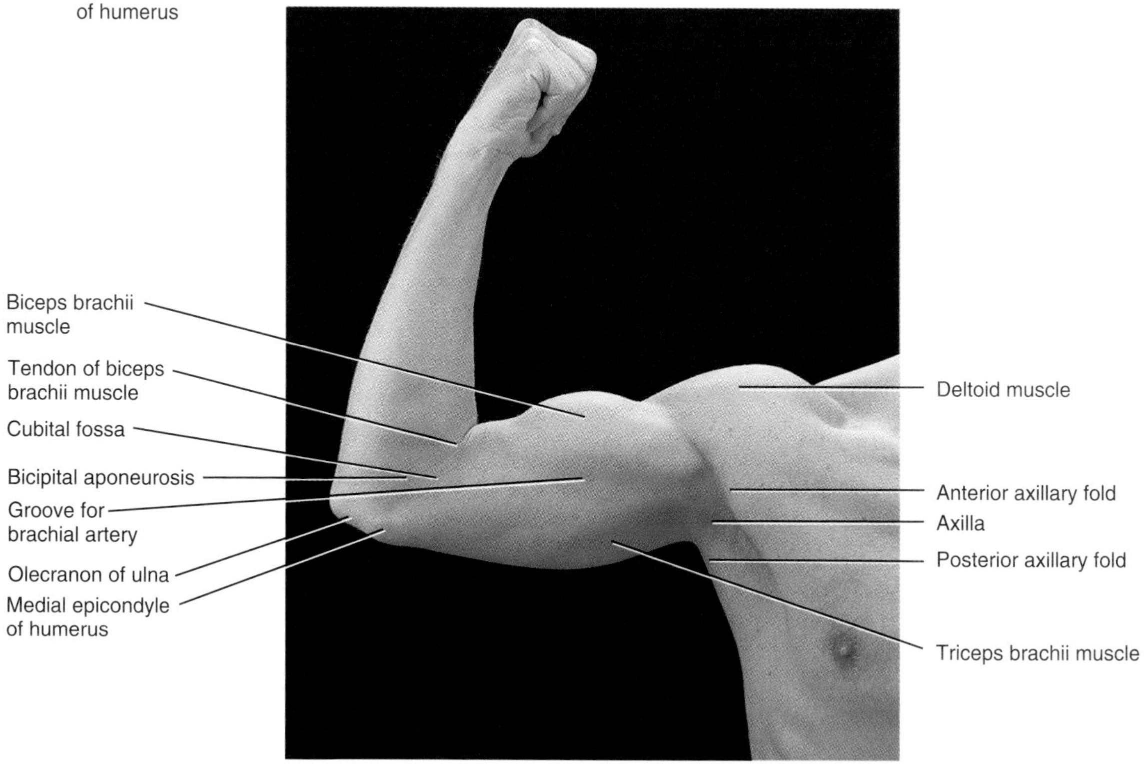

Biceps brachii muscle

Tendon of biceps brachii muscle

Cubital fossa

Bicipital aponeurosis

Groove for brachial artery

Olecranon of ulna

Medial epicondyle of humerus

Deltoid muscle

Anterior axillary fold

Axilla

Posterior axillary fold

Triceps brachii muscle

(b) Medial view of the arm and elbow

*Surface anatomy of the upper limb, continued*

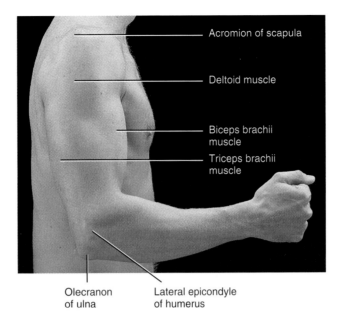

- Acromion of scapula
- Deltoid muscle
- Biceps brachii muscle
- Triceps brachii muscle
- Olecranon of ulna
- Lateral epicondyle of humerus

(c) Right lateral view of the arm and elbow

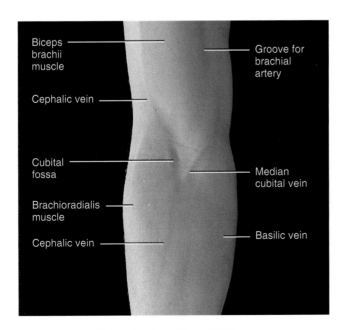

- Biceps brachii muscle
- Cephalic vein
- Cubital fossa
- Brachioradialis muscle
- Cephalic vein
- Groove for brachial artery
- Median cubital vein
- Basilic vein

(d) Anterior view of the cubital fossa

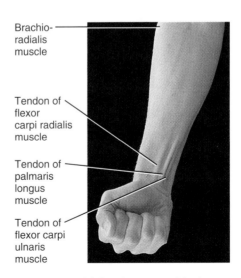

- Brachioradialis muscle
- Tendon of flexor carpi radialis muscle
- Tendon of palmaris longus muscle
- Tendon of flexor carpi ulnaris muscle

(e) Anterior aspect of the forearm and wrist

**205**

FIGURE 15.3 *Surface anatomy of the upper limb, continued* 206

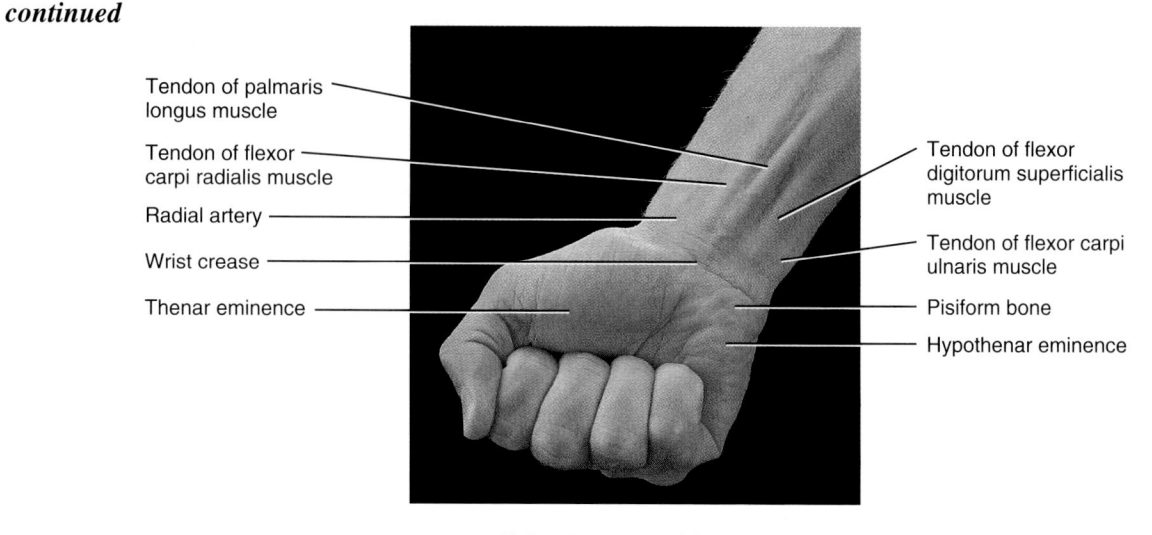

Tendon of palmaris longus muscle

Tendon of flexor carpi radialis muscle

Radial artery

Wrist crease

Thenar eminence

Tendon of flexor digitorum superficialis muscle

Tendon of flexor carpi ulnaris muscle

Pisiform bone

Hypothenar eminence

(f) Anterior aspect of the wrist

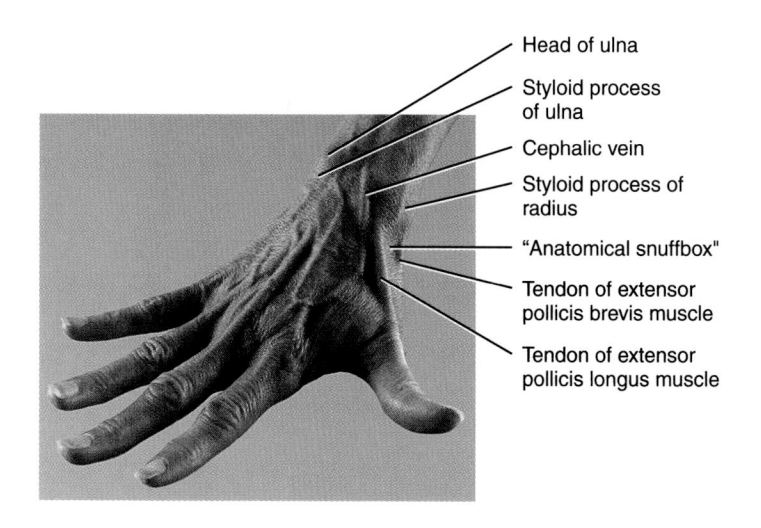

Head of ulna

Styloid process of ulna

Cephalic vein

Styloid process of radius

"Anatomical snuffbox"

Tendon of extensor pollicis brevis muscle

Tendon of extensor pollicis longus muscle

(g) Dorsum of the wrist

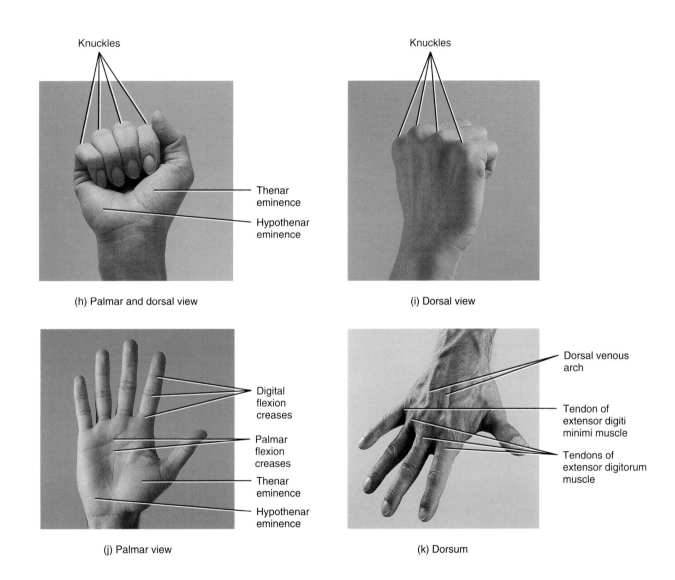

Knuckles

Thenar
eminence

Hypothenar
eminence

(h) Palmar and dorsal view

Knuckles

(i) Dorsal view

Digital
flexion
creases

Palmar
flexion
creases

Thenar
eminence

Hypothenar
eminence

(j) Palmar view

Dorsal venous
arch

Tendon of
extensor digiti
minimi muscle

Tendons of
extensor digitorum
muscle

(k) Dorsum

FIGURE 15.3 *Surface anatomy of the upper limb, continued*

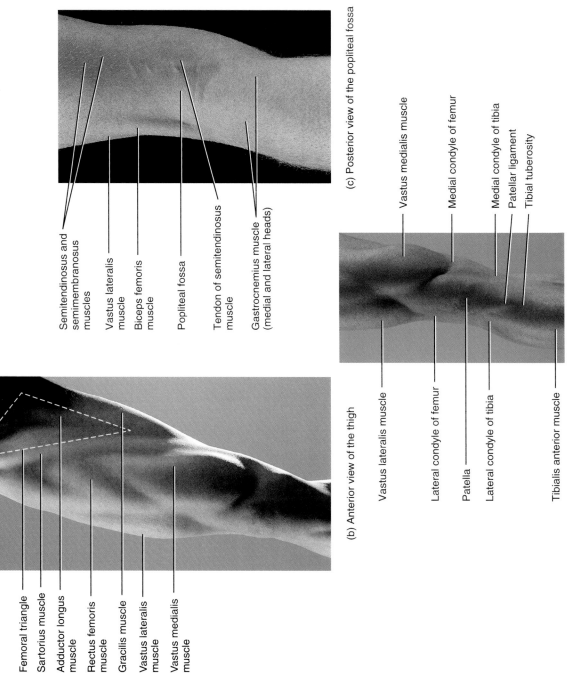

F I G U R E   1 5 . 4    |    *Surface anatomy of the lower limb*

Iliac crest

Gluteus medius muscle

Gluteus maximus muscle

Gluteal (natal) cleft

Greater trochanter

Ischial tuberosity

Gluteal fold

(a) Posterior view of the buttocks and thigh

Semitendinosus and semimembranosus muscles

Vastus lateralis muscle

Biceps femoris muscle

Popliteal fossa

Tendon of semitendinosus muscle

Gastrocnemius muscle (medial and lateral heads)

(c) Posterior view of the popliteal fossa

Vastus medialis muscle

Medial condyle of femur

Medial condyle of tibia

Patellar ligament

Tibial tuberosity

(d) Anterior view of the knee

Vastus lateralis muscle

Lateral condyle of femur

Patella

Lateral condyle of tibia

Tibialis anterior muscle

(b) Anterior view of the thigh

Femoral triangle

Sartorius muscle

Adductor longus muscle

Rectus femoris muscle

Gracilis muscle

Vastus lateralis muscle

Vastus medialis muscle

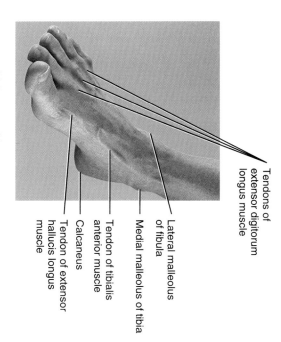

(g) Dorsum of foot

Tendons of extensor digitorum longus muscle

Lateral malleolus of fibula

Medial malleolus of tibia

Tendon of tibialis anterior muscle

Calcaneus

Tendon of extensor hallucis longus muscle

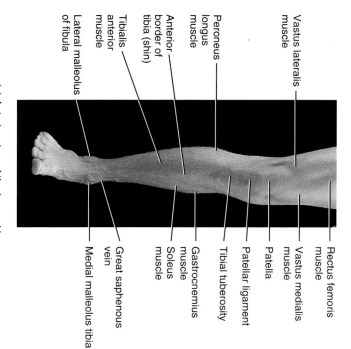

(e) Anterior view of the leg, ankle, and foot

Vastus lateralis muscle

Peroneus longus muscle

Anterior border of tibia (shin)

Tibialis anterior muscle

Lateral malleolus of fibula

Rectus femoris muscle

Vastus medialis muscle

Patella

Patellar ligament

Tibial tuberosity

Gastrocnemius muscle

Soleus muscle

Great saphenous vein

Medial malleolus tibia

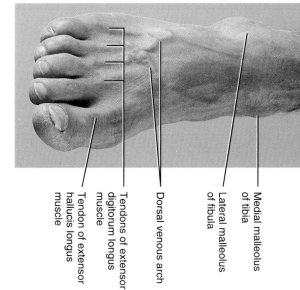

(h) Dorsum of foot

Tendons of extensor digitorum longus muscle

Dorsal venous arch

Lateral malleolus of fibula

Medial malleolus of tibia

Tendon of extensor hallucis longus muscle

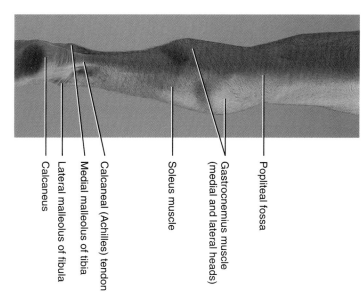

(f) Posterior view of the leg and ankle

Calcaneus

Lateral malleolus of fibula

Medial malleolus of tibia

Calcaneal (Achilles) tendon

Soleus muscle

Gastrocnemius muscle (medial and lateral heads)

Popliteal fossa

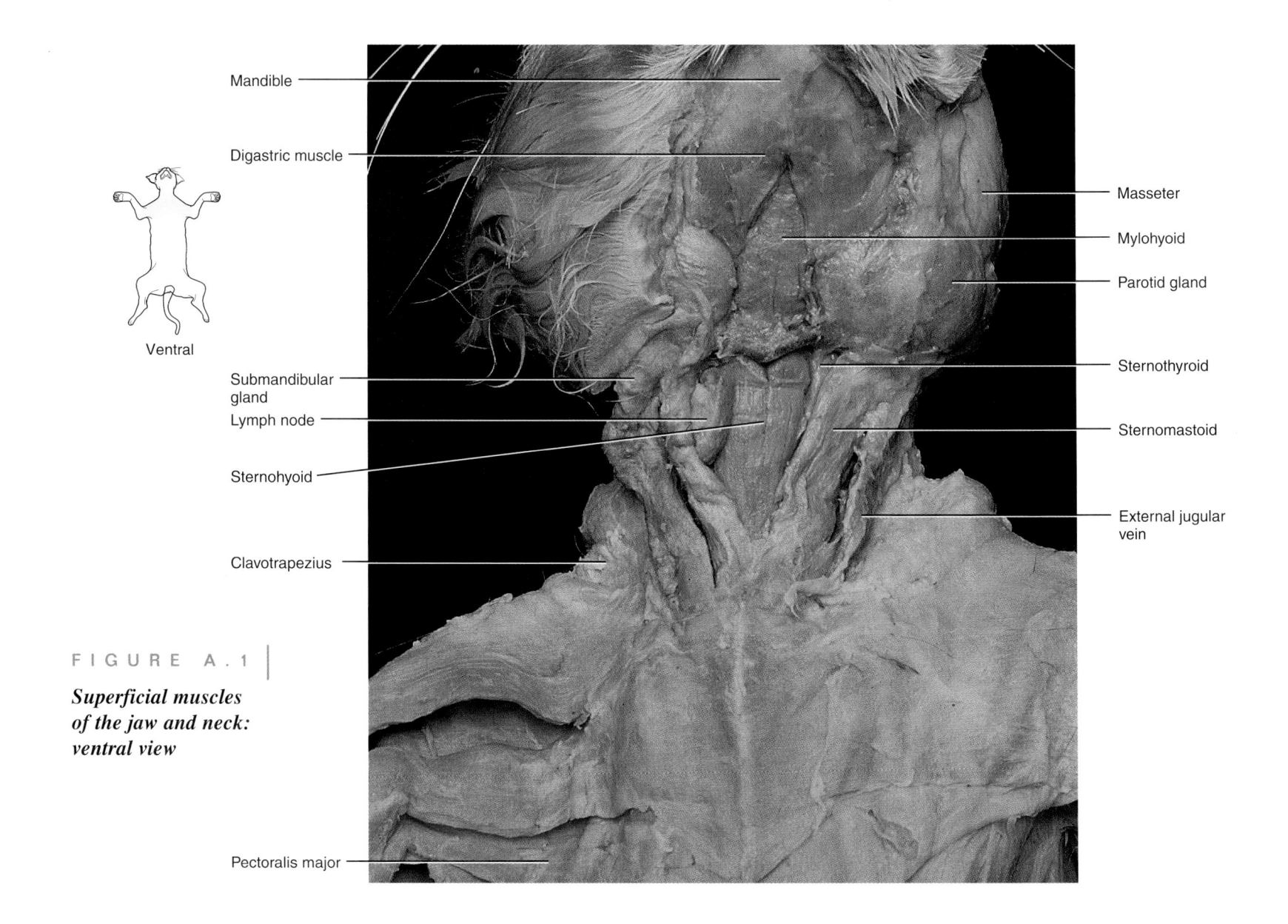

Mandible

Digastric muscle

Ventral

Masseter

Mylohyoid

Parotid gland

Submandibular
gland

Lymph node

Sternothyroid

Sternohyoid

Sternomastoid

External jugular
vein

Clavotrapezius

FIGURE A.1

*Superficial muscles
of the jaw and neck:
ventral view*

Pectoralis major

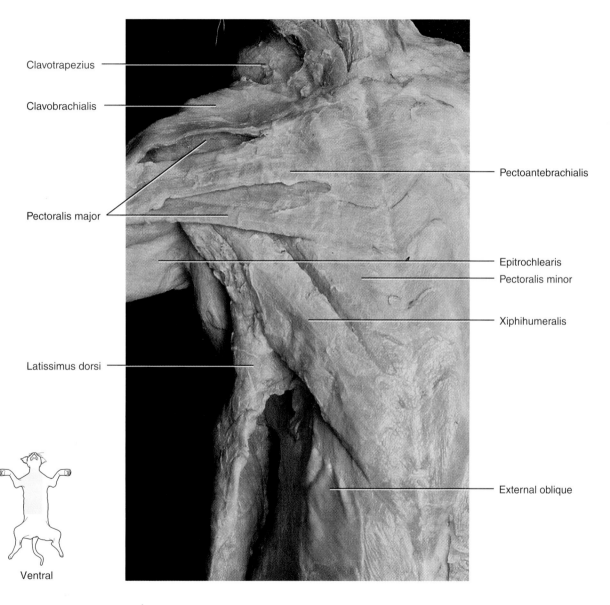

Clavotrapezius

Clavobrachialis

Pectoralis major

Latissimus dorsi

Ventral

Pectoantebrachialis

Epitrochlearis

Pectoralis minor

Xiphihumeralis

External oblique

FIGURE A.2a | *Superficial muscles of the chest: ventral view*

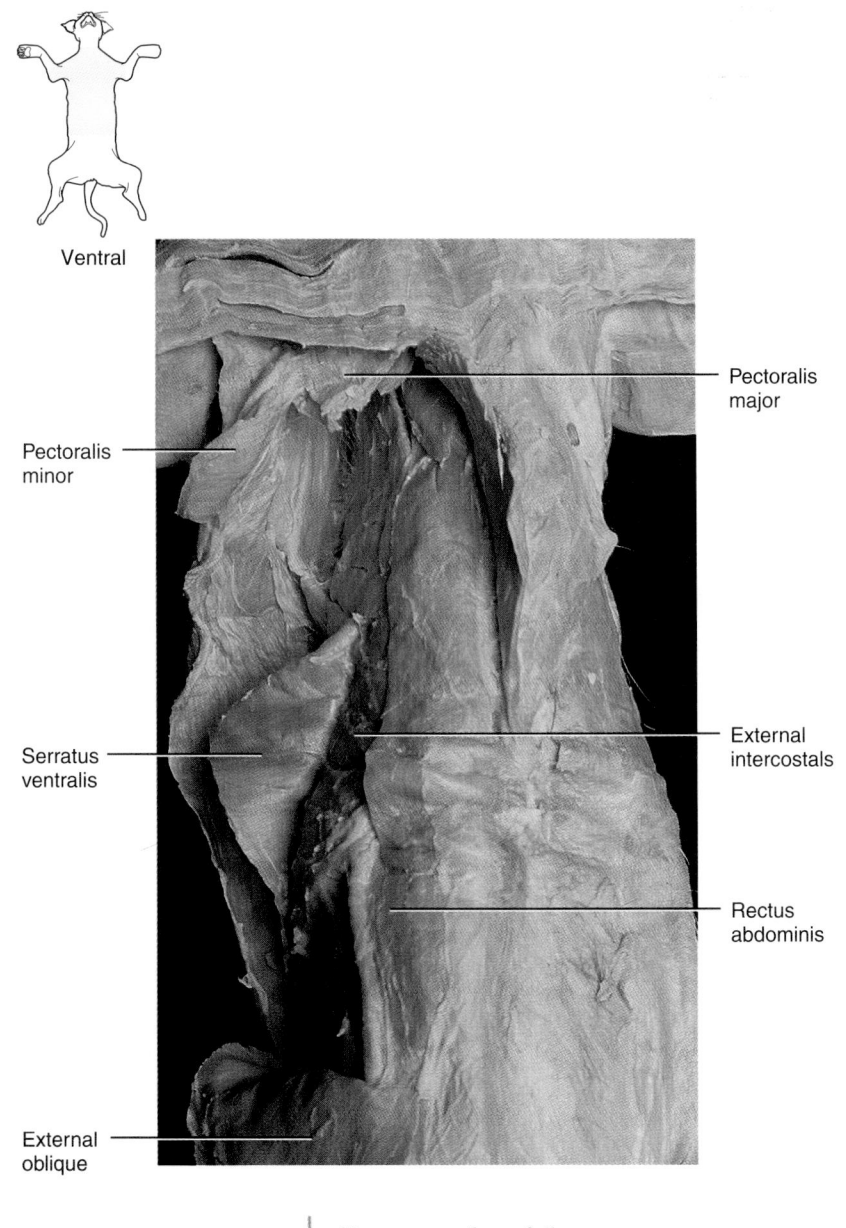

Ventral

Pectoralis
major

Pectoralis
minor

Serratus
ventralis

External
intercostals

Rectus
abdominis

External
oblique

FIGURE A.2b | *Deep muscles of the chest: ventral view*

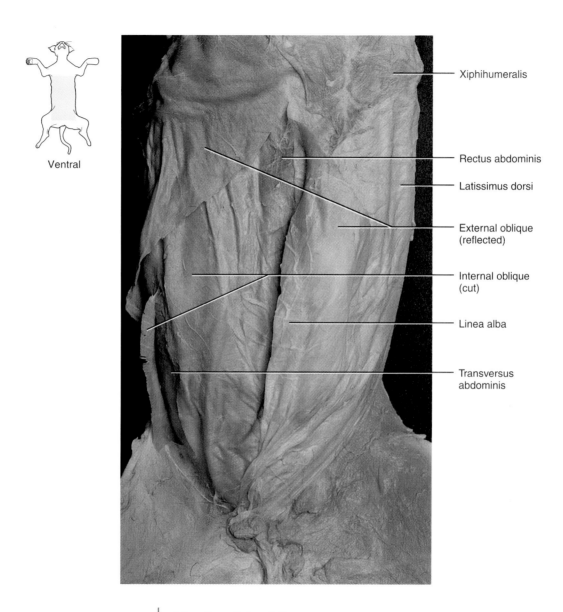

Ventral

Xiphihumeralis

Rectus abdominis

Latissimus dorsi

External oblique
(reflected)

Internal oblique
(cut)

Linea alba

Transversus
abdominis

FIGURE A.3 | *Muscles of the abdomen:*
*ventral view*

## Superficial muscles of the shoulder: left lateral view

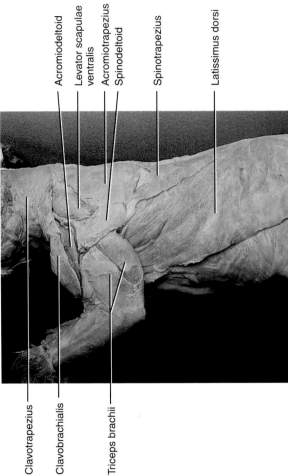

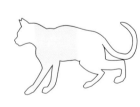

Left lateral

Acromiodeltoid
Levator scapulae ventralis
Acromiotrapezius
Spinodeltoid
Spinotrapezius
Latissimus dorsi

Clavotrapezius
Clavobrachialis
Triceps brachii

External oblique

## Deep muscles of the shoulder: left lateral view

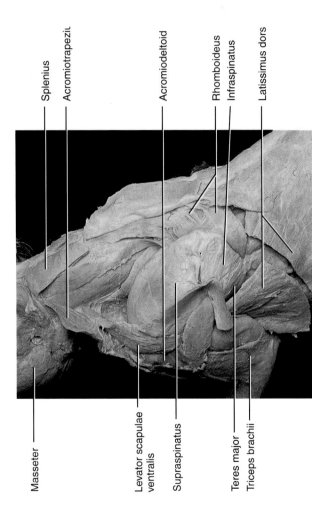

Left lateral

Splenius
Acromiotrapezit
Acromiodeltoid
Rhomboideus
Infraspinatus
Latissimus dors

Masseter
Levator scapulae ventralis
Supraspinatus
Teres major
Triceps brachii

Dorsal

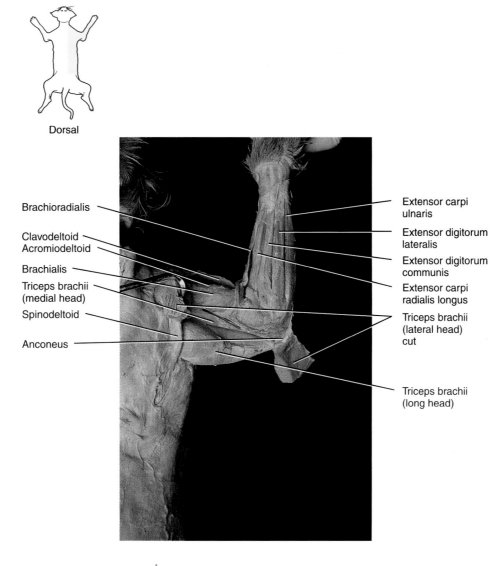

Brachioradialis

Clavodeltoid
Acromiodeltoid

Brachialis

Triceps brachii
(medial head)

Spinodeltoid

Anconeus

Extensor carpi
ulnaris

Extensor digitorum
lateralis

Extensor digitorum
communis

Extensor carpi
radialis longus

Triceps brachii
(lateral head)
cut

Triceps brachii
(long head)

FIGURE  A . 5  |  *Muscles of the arm and
forearm: lateral view*

Ventral

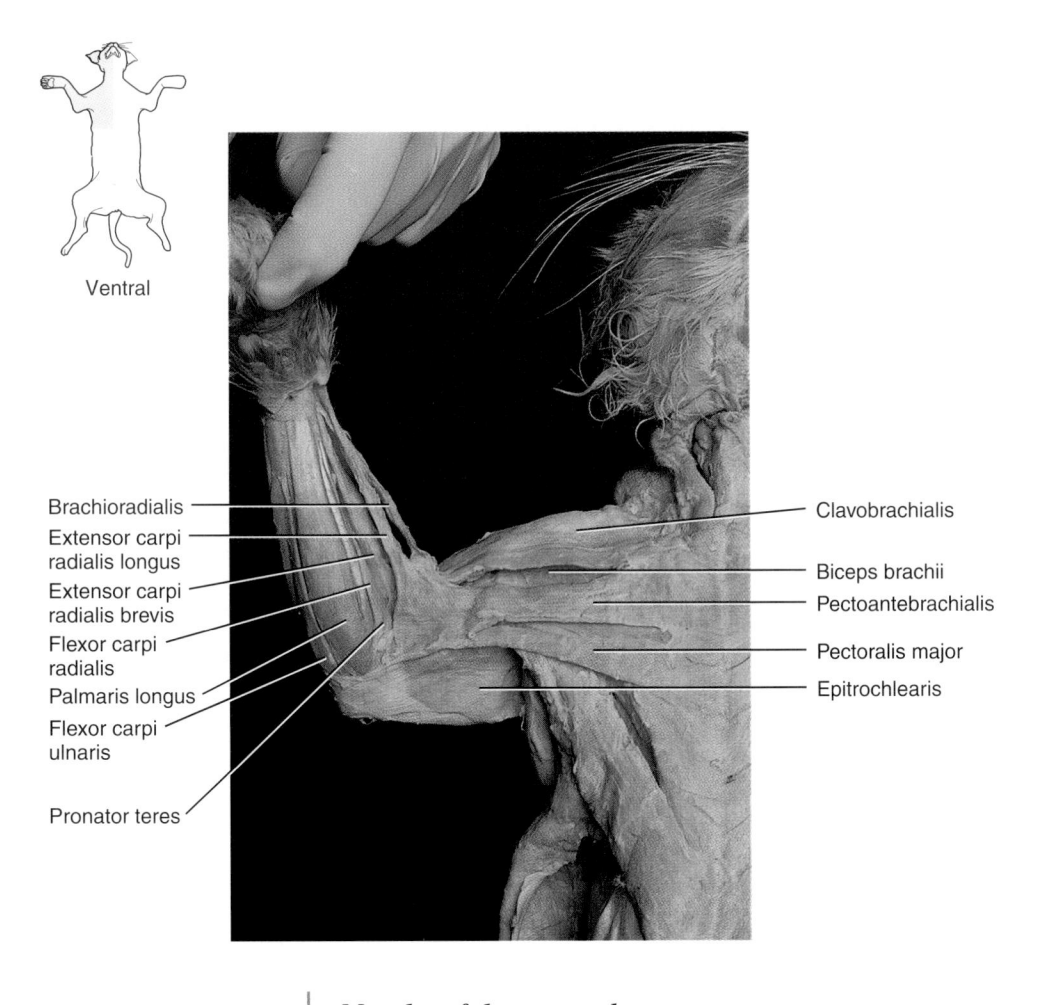

Brachioradialis

Extensor carpi
radialis longus

Extensor carpi
radialis brevis

Flexor carpi
radialis

Palmaris longus

Flexor carpi
ulnaris

Pronator teres

Clavobrachialis

Biceps brachii

Pectoantebrachialis

Pectoralis major

Epitrochlearis

FIGURE A.6 | *Muscles of the arm and
forearm: medial view*

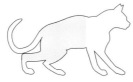

Right lateral

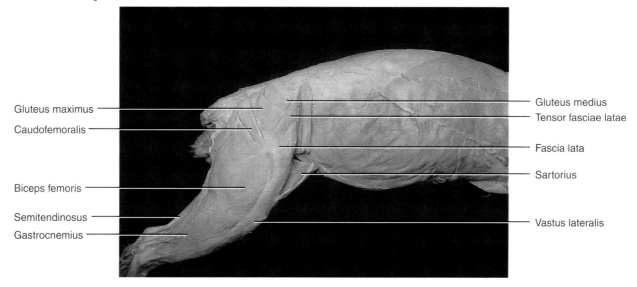

Gluteus maximus

Caudofemoralis

Biceps femoris

Semitendinosus

Gastrocnemius

Gluteus medius

Tensor fasciae latae

Fascia lata

Sartorius

Vastus lateralis

FIGURE A.7 | **Superficial muscles of the thigh: right lateral view**

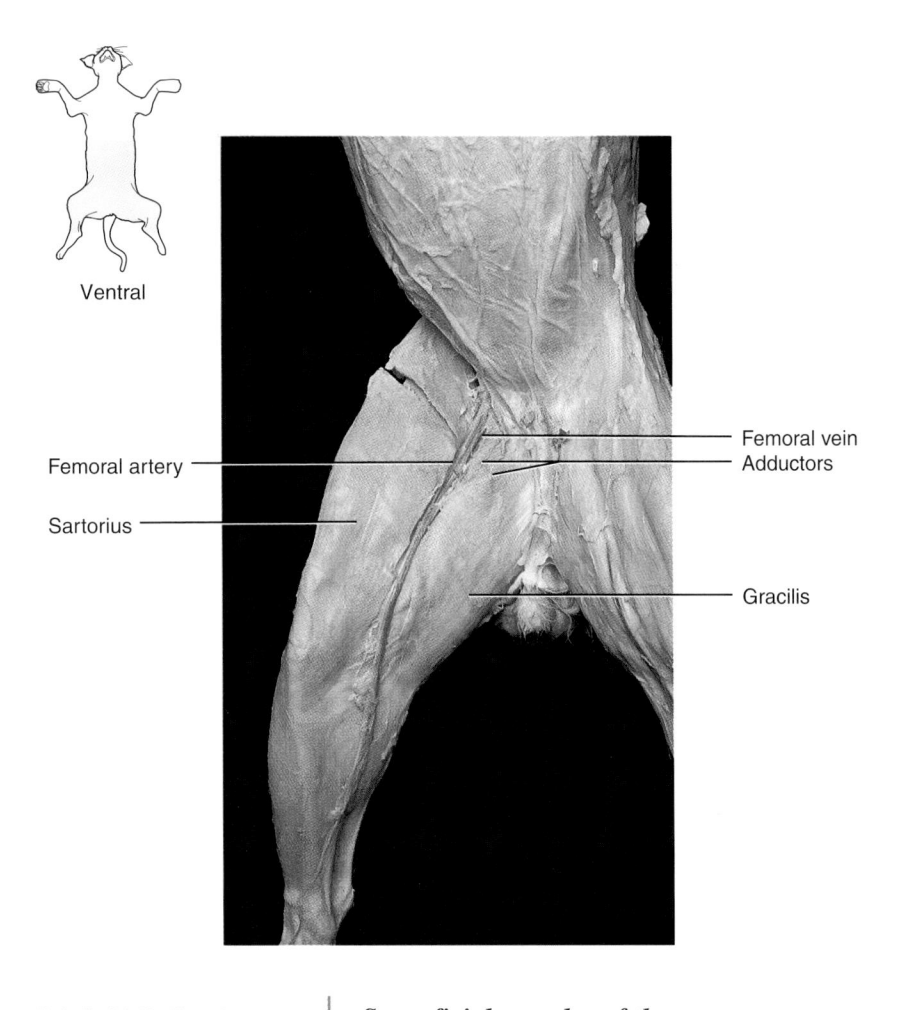

Ventral

Femoral artery

Sartorius

Femoral vein
Adductors

Gracilis

FIGURE A.8a | *Superficial muscles of the thigh: medial view*

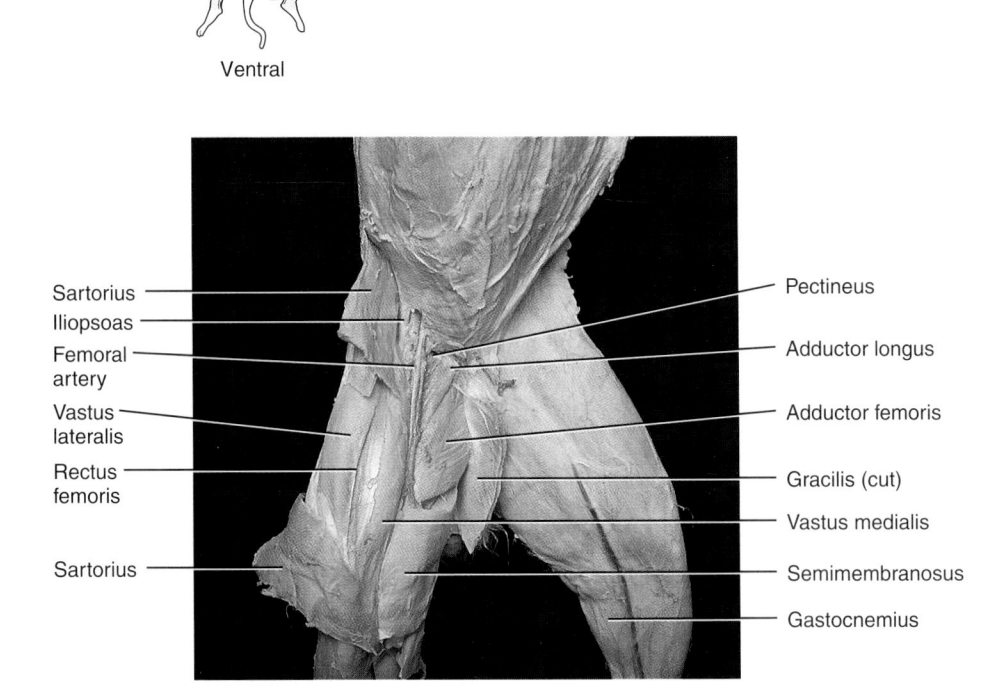

Ventral

Sartorius
Iliopsoas
Femoral artery
Vastus lateralis
Rectus femoris
Sartorius

Pectineus
Adductor longus
Adductor femoris
Gracilis (cut)
Vastus medialis
Semimembranosus
Gastocnemius

FIGURE A.8b | *Deep muscles of the thigh: medial view*

Right lateral

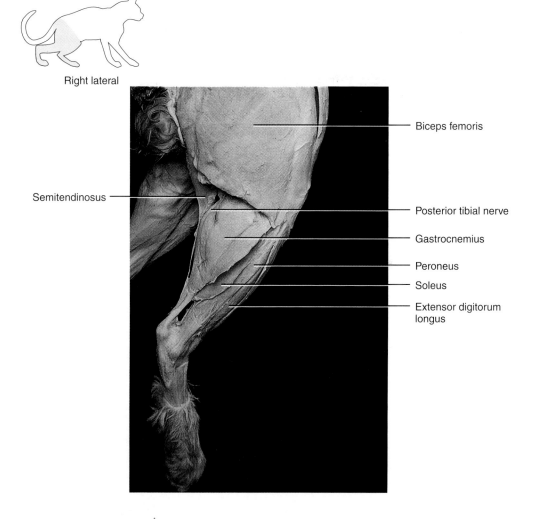

Biceps femoris

Semitendinosus

Posterior tibial nerve

Gastrocnemius

Peroneus

Soleus

Extensor digitorum
longus

FIGURE A.9 | *Superficial muscles of
the leg: right lateral view*

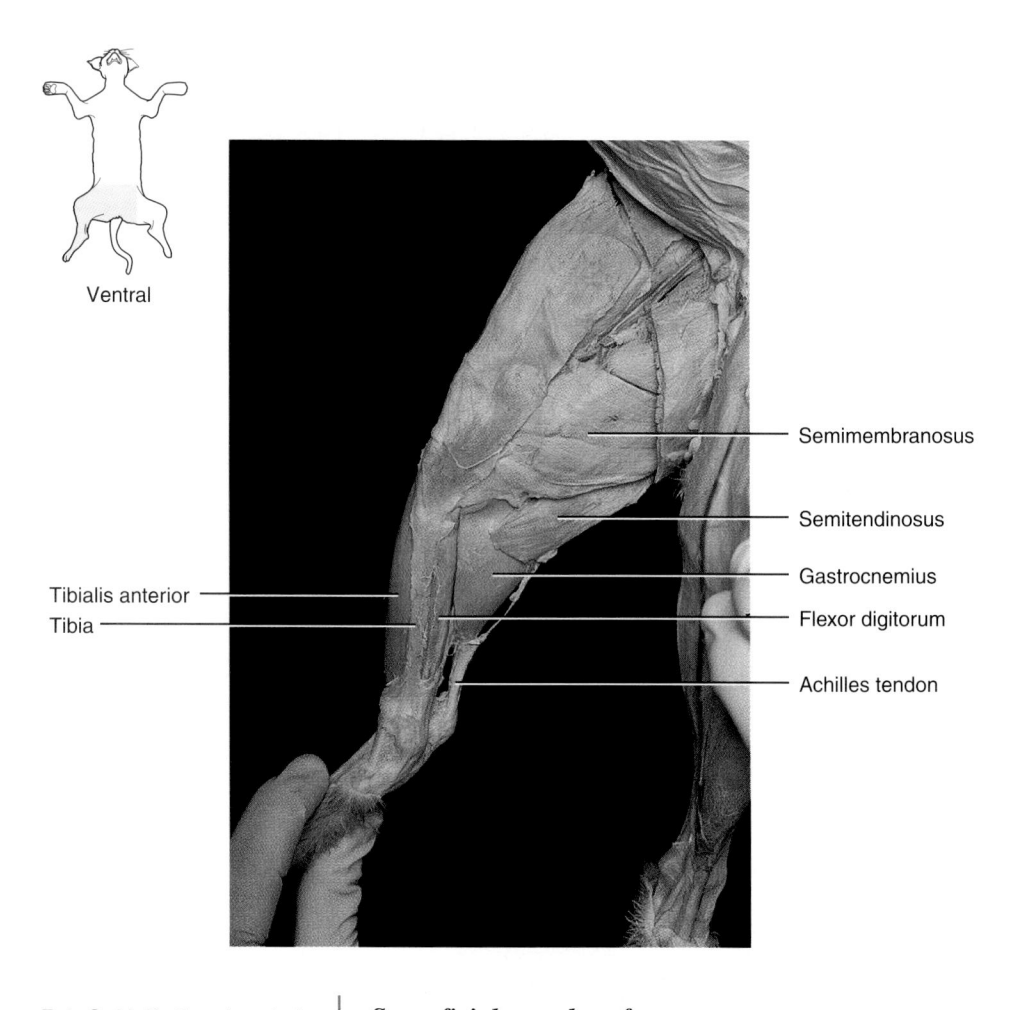

Ventral

Semimembranosus

Semitendinosus

Gastrocnemius

Tibialis anterior

Flexor digitorum

Tibia

Achilles tendon

FIGURE A.10 | *Superficial muscles of the leg: medial view*

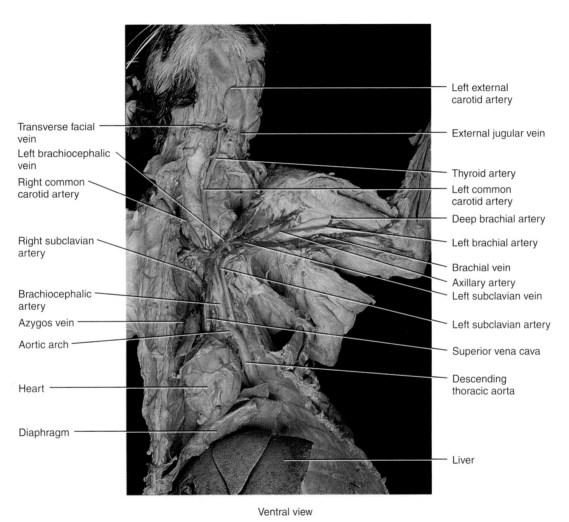

Left external
carotid artery

External jugular vein

Thyroid artery

Left common
carotid artery

Deep brachial artery

Left brachial artery

Brachial vein

Axillary artery

Left subclavian vein

Left subclavian artery

Superior vena cava

Descending
thoracic aorta

Liver

Transverse facial
vein

Left brachiocephalic
vein

Right common
carotid artery

Right subclavian
artery

Brachiocephalic
artery

Azygos vein

Aortic arch

Heart

Diaphragm

Ventral view

FIGURE A.11 | *Blood vessels above the diaphragm*

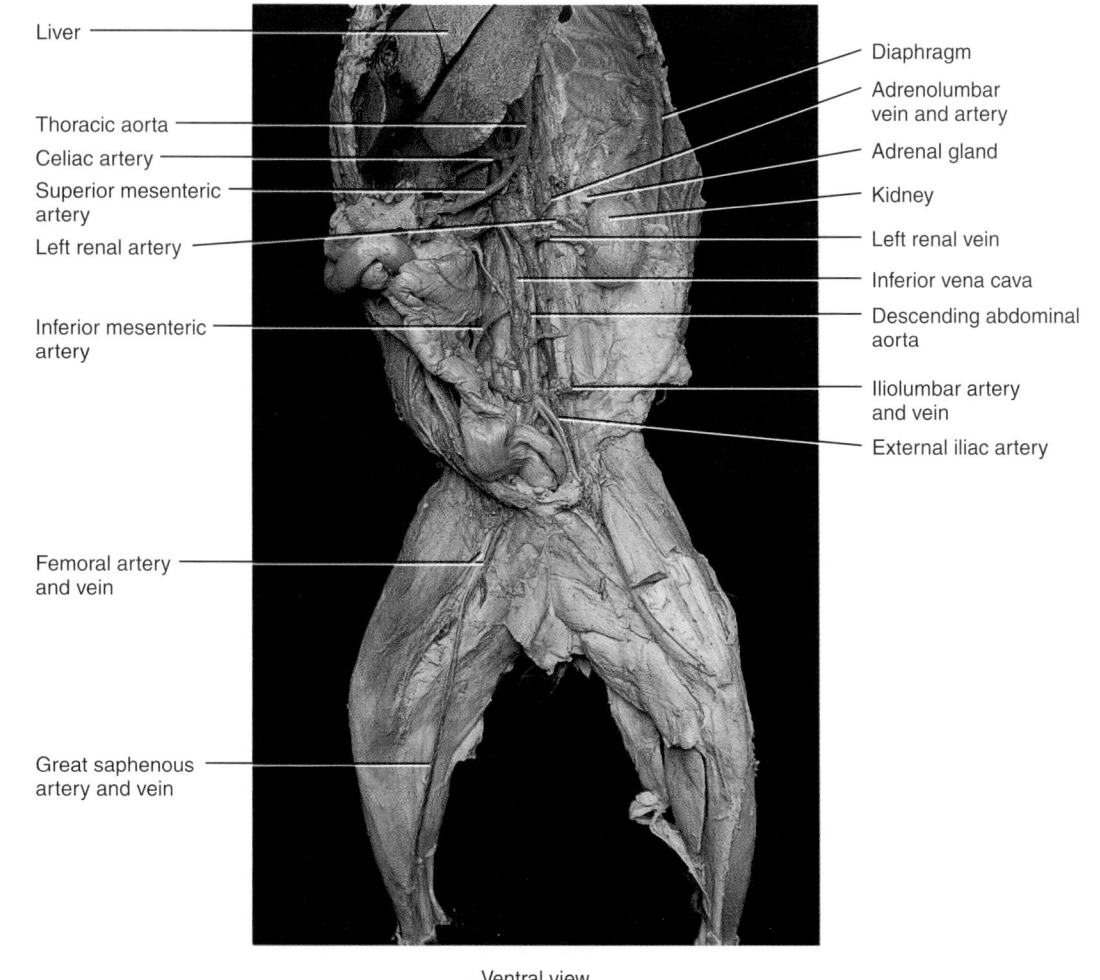

Liver

Thoracic aorta

Celiac artery

Superior mesenteric
artery

Left renal artery

Inferior mesenteric
artery

Femoral artery
and vein

Great saphenous
artery and vein

Diaphragm

Adrenolumbar
vein and artery

Adrenal gland

Kidney

Left renal vein

Inferior vena cava

Descending abdominal
aorta

Iliolumbar artery
and vein

External iliac artery

Ventral view

FIGURE A.12 | *Blood vessels below the diaphragm*

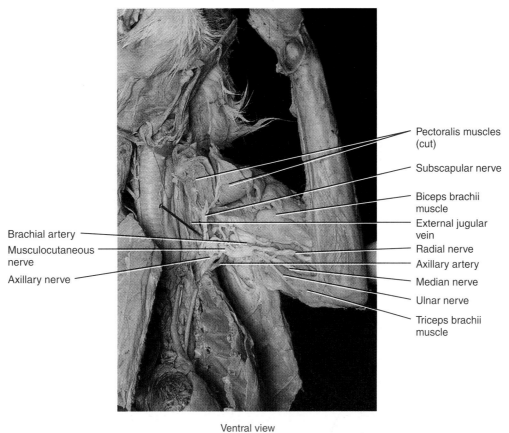

Pectoralis muscles (cut)

Subscapular nerve

Biceps brachii muscle

External jugular vein

Brachial artery

Radial nerve

Musculocutaneous nerve

Axillary artery

Median nerve

Axillary nerve

Ulnar nerve

Triceps brachii muscle

Ventral view

FIGURE A.13 | *Brachial plexus and associated blood vessels*

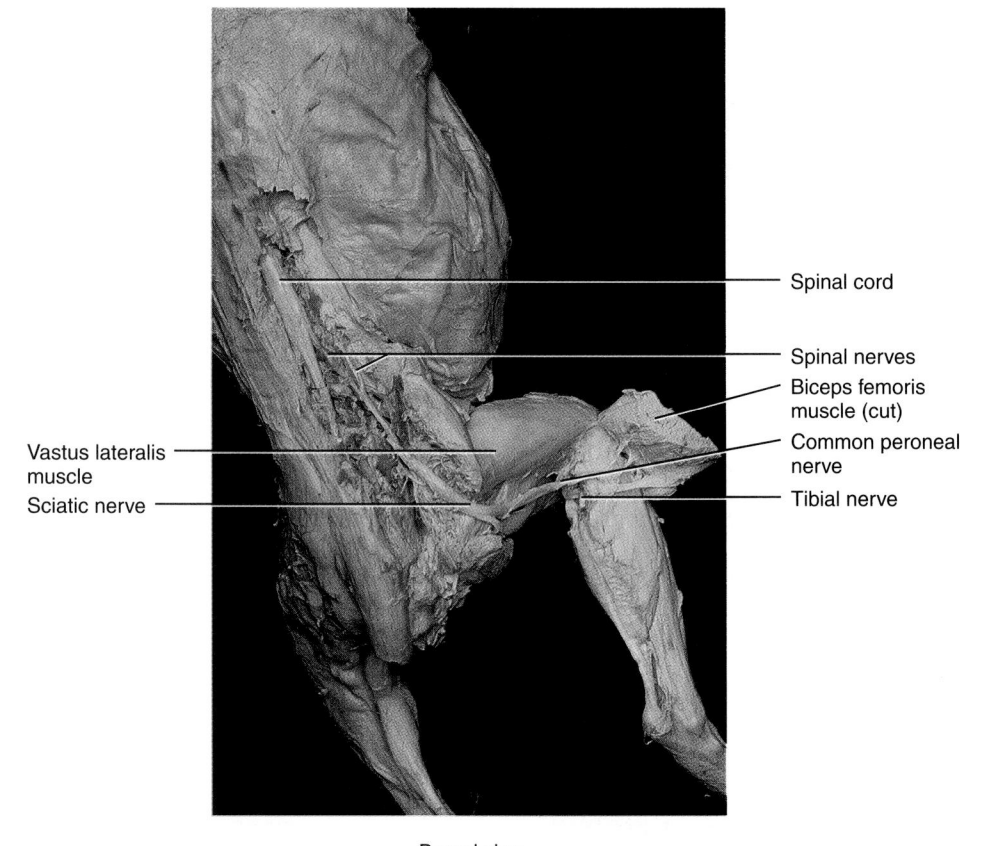

Spinal cord

Spinal nerves

Biceps femoris
muscle (cut)

Common peroneal
nerve

Tibial nerve

Vastus lateralis
muscle

Sciatic nerve

Dorsal view

FIGURE A.14 | *Spinal cord and sacral plexus*

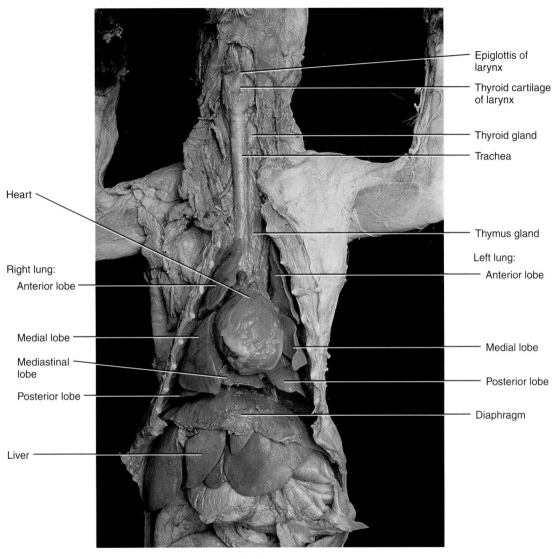

Epiglottis of
larynx

Thyroid cartilage
of larynx

Thyroid gland

Trachea

Heart

Thymus gland

Right lung:

Left lung:

Anterior lobe

Anterior lobe

Medial lobe

Medial lobe

Mediastinal
lobe

Posterior lobe

Posterior lobe

Diaphragm

Liver

Ventral view

FIGURE  A . 1 5  |  *Respiratory system*

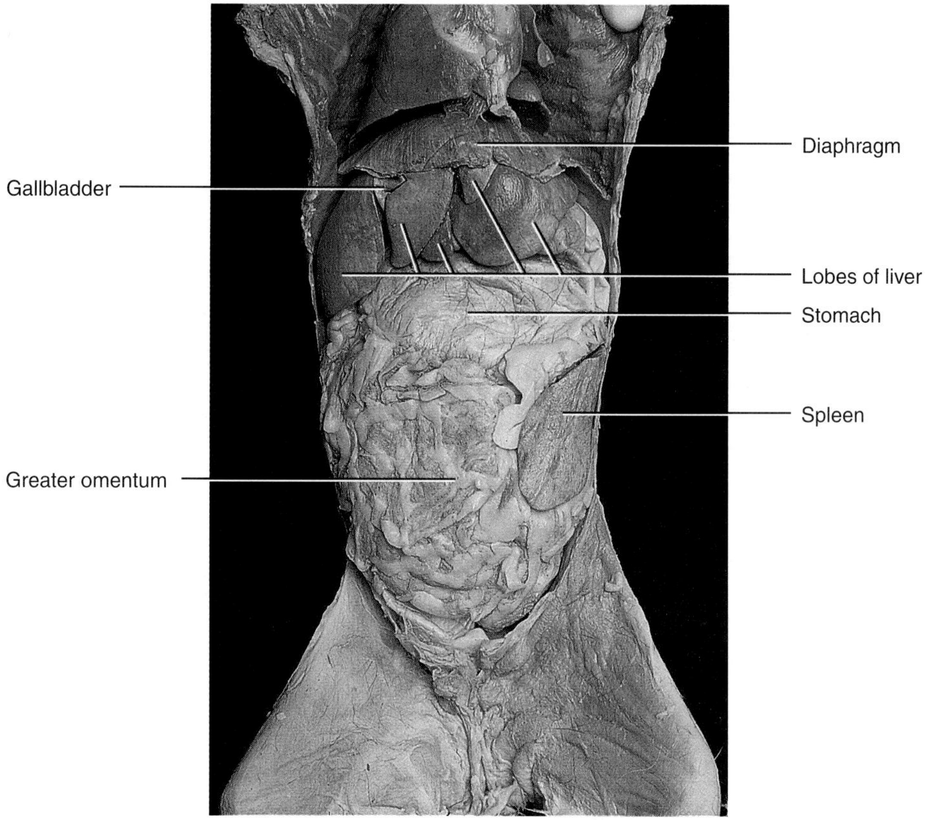

Gallbladder

Diaphragm

Lobes of liver

Stomach

Spleen

Greater omentum

Ventral view

FIGURE A.16a | *Digestive system, superficial organs*

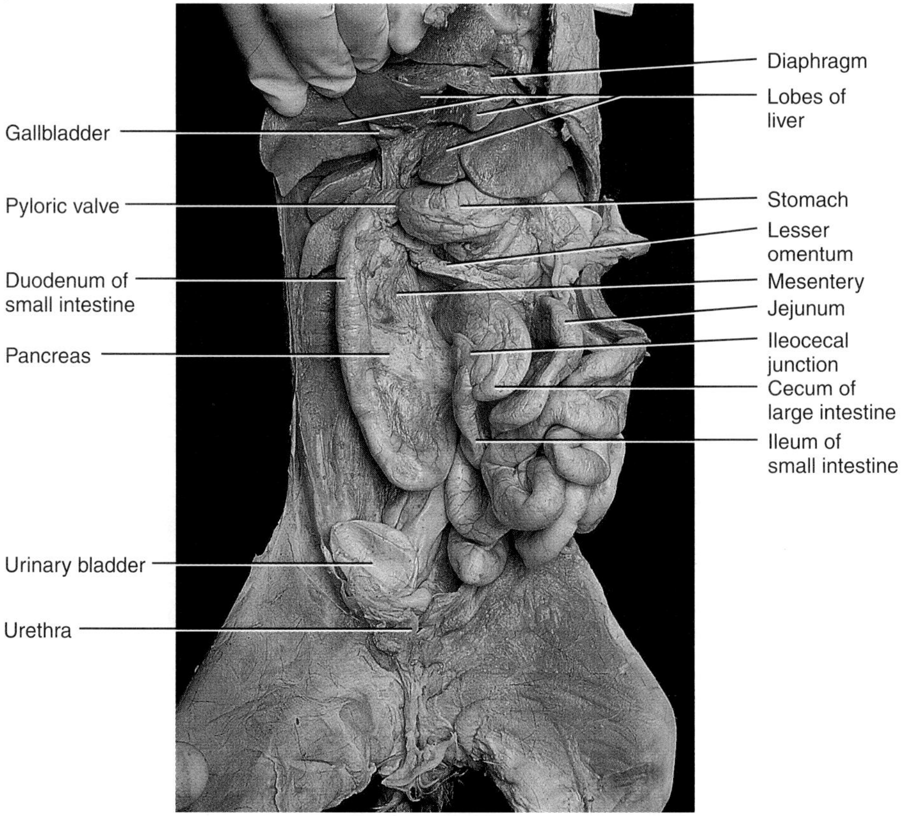

Gallbladder

Pyloric valve

Duodenum of small intestine

Pancreas

Urinary bladder

Urethra

Diaphragm

Lobes of liver

Stomach

Lesser omentum

Mesentery

Jejunum

Ileocecal junction

Cecum of large intestine

Ileum of small intestine

Ventral view

FIGURE A.16b | *Digestive system, deep organs*

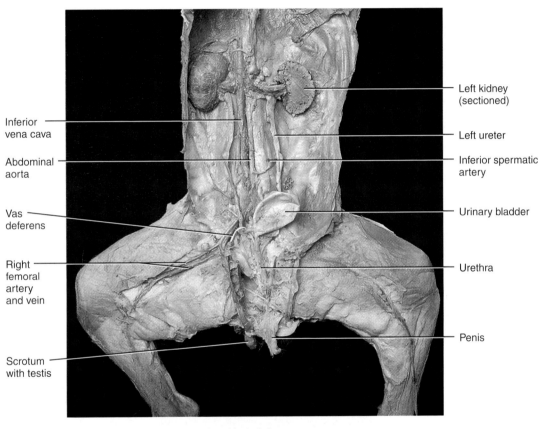

Inferior
vena cava

Abdominal
aorta

Vas
deferens

Right
femoral
artery
and vein

Scrotum
with testis

Left kidney
(sectioned)

Left ureter

Inferior spermatic
artery

Urinary bladder

Urethra

Penis

Ventral view

FIGURE A.17a | *Male urinary system*

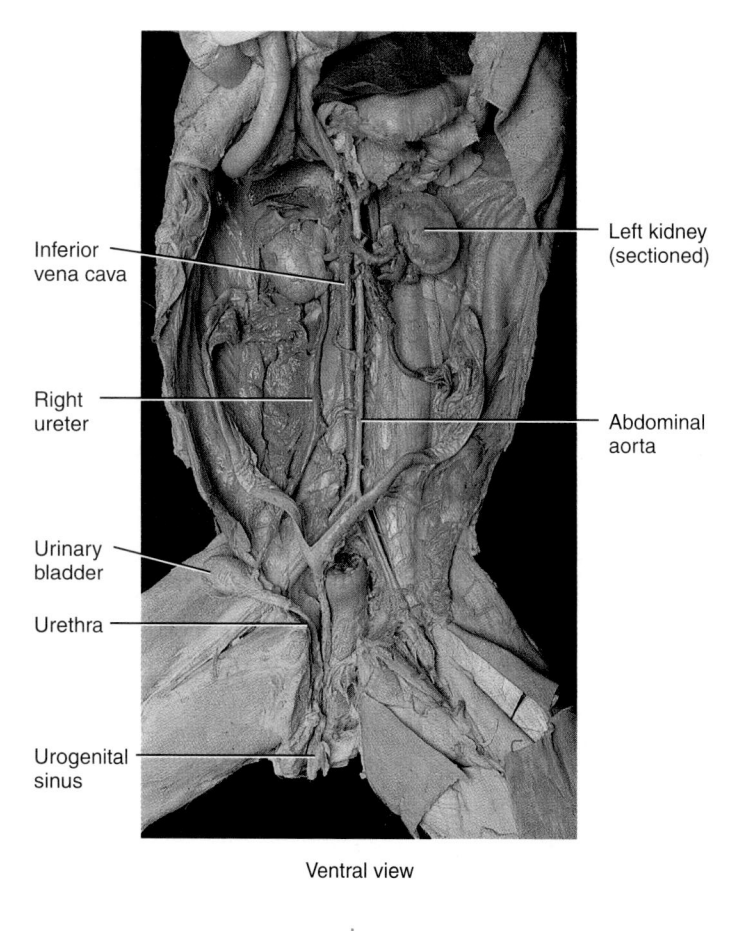

Inferior
vena cava

Left kidney
(sectioned)

Right
ureter

Abdominal
aorta

Urinary
bladder

Urethra

Urogenital
sinus

Ventral view

FIGURE  A. 17 b  |  *Female urinary system*

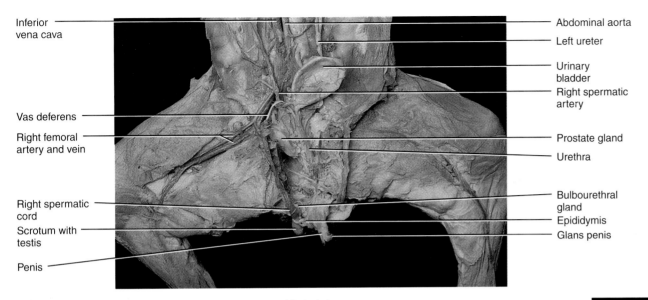

Inferior vena cava

Vas deferens

Right femoral artery and vein

Right spermatic cord

Scrotum with testis

Penis

Abdominal aorta

Left ureter

Urinary bladder

Right spermatic artery

Prostate gland

Urethra

Bulbourethral gland

Epididymis

Glans penis

Ventral view

FIGURE  A . 1 8 a  |  *Male reproductive system*

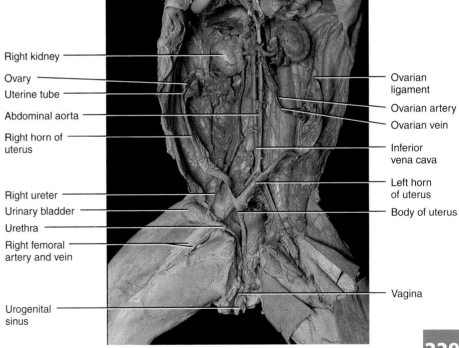

Right kidney

Ovary

Uterine tube

Abdominal aorta

Right horn of uterus

Right ureter

Urinary bladder

Urethra

Right femoral artery and vein

Urogenital sinus

Ovarian ligament

Ovarian artery

Ovarian vein

Inferior vena cava

Left horn of uterus

Body of uterus

Vagina

FIGURE  A . 1 8 b  |  *Female reproductive system*

Ventral view

# CROSS REFERENCE GUIDE TO

# A.D.A.M.® INTERACTIVE ANATOMY

## for *A Photographic Atlas of the Human Body with Selected Cat, Sheep, and Cow Dissections*, by Gerard J. Tortora

**Prepared by Barbara Stewart, Burlington County College**

This reference guide, which follows the organization of your atlas, directs you to links between the illustrations in the atlas and multiple views, including 3D images, cadaver photographs, and pinned atlas images. To begin, start A.D.A.M.® Interactive Anatomy by clicking the appropriate icon in the program list. Then follow the instructions below.

**Working with A.D.A.M.® Interactive Anatomy:**

Dissectible Anatomy:

1. In the open dialog window, select the Gender and View in the guide and click Open.

2. Scroll to the layer indicated under Scroll To in the guide.

3. The image corresponding to the image in the photo atlas appears.

4. Click on any structure in the image frame and its label appears as pop-up text.

Atlas Anatomy:

1. In the open dialog window select the appropriate View and System option from the list of views and systems under Show Images For. You will find the view listed in the View/Show Images For column, and the system under Region/System/View/Type/All in the guide.

2. Find and select the Image Title from the Image List and click Open.

3. Click on any visible pins and the name of the pinned structure appears as pop-up text.

3D Anatomy:

1. In the open dialog window select the structure you want to view, which is listed under View/Show Images For in the guide and click Open.

2. In the drop-down menu at the top of the window, scroll to the structure listed in the Scroll To column in the guide.

Good luck with your anatomy studies!

*Barbara Stewart*

| Atlas Figure | Section | Gender | View/Show Images For | Region/ System/View/ Type/All | Scroll To | Image Title |
|---|---|---|---|---|---|---|
| 3.2 | Dissectible | | Anterior | | Layer # 48 | |
| 3.3 | Atlas | | System | Skeletal | | Skull (Lat) 1 |
| 3.4 | Atlas | | View | Medial | | Muscle Atts-Skull (Med) |
| 3.5 | Dissectible | | Posterior | | Layer # 14 | |
| 3.6 | Atlas | | Region | Head & Neck | | Skull (Lat) 1 |
| 3.7 | Atlas | | Region | Head & Neck | | Skull (Inf) |
| 3.8 | Atlas | | Region | Head & Neck | | Cranial Cavities (Sup) 1 |
| 3.9 a, b, c | 3D | | 3D Skull | | Sphenoid Bone - Isolated | |
| 3.10 a | Atlas | | Region | Head & Neck | | Muscle Atts-Skull (Med) |
| 3.10 b, c, d | 3D | | 3D Skull | | Ethmoid Bone -Isolated | |
| 3.11 | Atlas | | System | Skeletal | | Walls of Orbit (Ant) |
| 3.13 | Atlas | | View | Superior | | Laryngeal Muscles (Sup) |
| 3.14 | Atlas | | Region | Head & Neck | | Ear Ossicles (Med) |
| 3.15 | Atlas | | View | Lateral | | Bones of Trunk (Lat) |
| 3.16 a, b | Atlas | | Type | Cadaver Photograph | | Isolated Vertebrae (Ant/Lat) |
| 3.17 a | Dissectible | | Anterior | | Layer # 329 | |
| 3.17 b, c | Dissectible | | Anterior | | Layer # 330 | |
| 3.18 a, b | Atlas | | Type | Cadaver Photograph | | Lumbar Vertebrae (Post/Lat) |
| 3.19 a | Atlas | | Type | Cadaver Photograph | | Dorsal Surface of Sacrum |
| 3.19 b | Atlas | | Type | Cadaver Photograph | | Pelvic Surface of Sacrum |
| 3.20 | Atlas | | Type | Cadaver Photograph | | Dissection of Thorax (Ant) |
| 3.21 a, b | Dissectible | | Posterior | | Layer # 176 | |
| 3.22 a | Atlas | | View | Anterior | | Bones of Arm & Shoulder (Ant) |
| 3.22 b | Dissectible | | Posterior | | Layer # 71 | |
| 3.22 c | Dissectible | | Lateral | | Layer # 38 | |
| 3.23 a | Atlas | | Type | Illustration | | Bones of Arm & Shoulder (Ant) |

| Atlas Figure | Section | Gender | View/Show Images For | Region/ System/View/ Type/All | Scroll To | Image Title |
|---|---|---|---|---|---|---|
| 3.23 b | Atlas | | Type | Illustration | | Bones of Arm & Shoulder (Post) |
| 3.24 a | Atlas | | Type | Illustration | | Bones of Forearm & Hand (Ant) |
| 3.24 b | Atlas | | Type | Illustration | | Bones of Forearm & Hand (Post) |
| 3.25 | Atlas | | Type | Illustration | | Bones of Hand (Ant) |
| 3.26 a | Atlas | | View | Lateral | | Male Bony Pelvis (Lat) |
| 3.26 b | Dissectible | | Medial | | Layer # 127 | |
| 3.27 a | Atlas | | View | Superior | | Female Bony Pelvis (Sup) |
| 3.27 b | Atlas | | View | Superior | | Male Bony Pelvis (Sup) |
| 3.28 a | Dissectible | | View | Anterior | Layer # 329 | |
| 3.28 b | Dissectible | | View | Posterior | Layer # 185 | |
| 3.29 | Dissectible | | View | Anterior | Layer # 330 | |
| 3.30 a | Dissectible | | View | Anterior | Layer # 329 | |
| 3.30 b | Dissectible | | View | Posterior | Layer # 185 | |
| 3.31 | Atlas | | Type | Illustration | | Bones of Foot (Dorsal) |

UNIT 4 | *Articulations*

| Atlas Figure | Section | Gender | View/Show Images For | Region/ System/View/ Type/All | Scroll To | Image Title |
|---|---|---|---|---|---|---|
| 4.1 | Atlas | | Type | Cadaver Photograph | | Brachial Plexus |
| 4.2 | Dissectible | | Anterior | | Layer # 329 | |
| 4.3 | Dissectible | | Anterior | | Layer # 142 | |
| 4.4 | Dissectible | | Lateral Arm | | Layer # 64 | |
| 4.5 | Dissectible | | Anterior | | Layer # 330 | |

| Atlas Figure | Section | Gender | View/Show Images For | Region/ System/View/ Type/All | Scroll To | Image Title |
|---|---|---|---|---|---|---|
| 4.6 | Dissectible | | Anterior | | Layer # 315 | |
| 4.7 | Dissectible | | Posterior | | Layer # 168 | |
| 4.8 | Dissectible | | Medial | | Layer # 120 | |

U N I T   5   | *The Muscular System*

| Atlas Figure | Section | Gender | View/Show Images For | Region/ System/View/ Type/All | Scroll To | Image Title |
|---|---|---|---|---|---|---|
| 5.2 | Dissectible | | Anterior | | Layer # 13 | |
| 5.3 | Dissectible | | Lateral | | Layer # 39 | |
| 5.4 | Dissectible | | Lateral | | Layer # 36 | |
| 5.5 | Atlas | | View | Lateral | | Extrinsic Eye Muscles (Lat) |
| 5.6 | Atlas | | Type | Cadaver Photograph | | Muscle Atts-Larynx (Post) |
| 5.7 | Dissectible | | Anterior | | Layer # 29 | |
| 5.8 | Dissectible | | Lateral | | Layer # 8 | |
| 5.9 | Dissectible | | Posterior | | Layer # 10 | |
| 5.10 | Dissectible | | Posterior | | Layer # 11 | |
| 5.11 | Dissectible | | Anterior | | Layer # 19 | |
| 5.12 | Dissectible | | Anterior | | Layer # 103 | |
| 5.13 | Dissectible | | Posterior | | Layer # 23 | |
| 5.14 | Dissectible | | Anterior | | Layer # 117 | |
| 5.15 | Dissectible | | Posterior | | Layer # 25 | |
| 5.16 | Dissectible | | Anterior | | Layer # 182 | |
| 5.17 | Dissectible | | Posterior | | Layer # 83 | |
| 5.18 | Atlas | | Type | Cadaver Photograph | | Anterior Leg |
| 5.19 | Dissectible | | Posterior | | Layer # 137 | |

| Atlas Figure | Section | Gender | View/Show Images For | Region/ System/View/ Type/All | Scroll To | Image Title |
|---|---|---|---|---|---|---|
| 6.3 | 3D | | 3D Heart | | | Heart |
| 6.4 | Dissectible | | Anterior | | Layer # 174 | |
| 6.5 | Atlas | | Type | Illustration | | Right Mediastinum (Lat) |
| 6.6 | Atlas | | View | Superior | | Mediastinum (Sup) |
| 6.7 | Atlas | | View | Superior | | Coronary Arteries (Sup) |
| 6.8 | Atlas | | System | Cardiovascular | | Heart and Great Vessels (Ant) |
| 6.9 | 3D | | 3D Heart | | | Coronary Arteries (Ant) |

UNIT 7 | *The Lympathic System*

| Atlas Figure | Section | Gender | View/Show Images For | Region/ System/View/ Type/All | Scroll To | Image Title |
|---|---|---|---|---|---|---|
| 7.2 | Atlas | | Type | Cadaver Photograph | | Dissection of the Left Mediastinum |
| 7.3 | Dissectible | | | Anterior | | Layer # 214 |
| 7.4 | Atlas | | System | Lymphatic | | Lymph Nodes of the Thorax (Ant) |

| Atlas Figure | Section | Gender | View/Show Images For | Region/ System/View/ Type/All | Scroll To | Image Title |
|---|---|---|---|---|---|---|
| 8.1a | Atlas | | View | Medial | | Sagittal Section of Brain |
| 8.1b | Atlas | | System | Nervous | | Spinal Cord Vessels and Meninges |
| 8.2 | Atlas | | System | Nervous | | Arteries of Spinal Cord (Post) |
| 8.3 | Atlas | | System | Nervous | | Lumbosacral Spinal Cord (Post) |
| 8.4 | Atlas | | System | Nervous | | T12 Vertebrae (Sup) |
| 8.6 | Atlas | | System | Nervous | | Arteries of Spinal Cord (Post) |
| 8.7 | Dissectible | | Lateral | | Layer # 17 | |
| 8.8 | Atlas | | System | Nervous | | Brachial Plexus |
| 8.9 | Atlas | | Type | Cadaver Photograph | | Anterior Thoracic Wall (Post) |
| 8.10 | Dissectible | | Anterior | | Layer # 217 | |
| 8.11 | Atlas | | System | Nervous | | Sacral Plexus (Med) |
| 8.12 | Dissectible | | Lateral | | Layer # 191 | |
| 8.13 | Atlas | | Region | Head and Neck | | Sagittal Section of Brain |
| 8.14 | Atlas | | System | Nervous | | Brain (Lat) |
| 8.15 | Dissectible | | Lateral | | Layer # 288 | |
| 8.18 | Dissectible | | Anterior | | Layer # 275 | |
| 8.19 | Atlas | | System | Nervous | | Base of Brain (Inf) |
| 8.21 | Atlas | | View | Superior | | Basal Ganglia (Sup) |

| Atlas Figure | Section | Gender | View/Show Images For | Region/ System/View/ Type/All | Scroll To | Image Title |
|---|---|---|---|---|---|---|
| 9.1b | Atlas | | System | Respiratory | | Lateral Wall of Nasal Cavity |
| 9.2 | Atlas | | System | Respiratory | | Dorsal Surface of Tongue |
| 9.3b | Atlas | | System | Nervous | | Sagittal Section of Eyeball |

UNIT 10 | *The Endocrine System*

| Atlas Figure | Section | Gender | View/Show Images For | Region/ System/View/ Type/All | Scroll To | Image Title |
|---|---|---|---|---|---|---|
| 10.2 | Atlas | | Region | Head and Neck | | Glands of the Head & Neck (Lat) |
| 10.3 | Dissectible | | Anterior | | Layer # 78 | |
| 10.5a | Dissectible | | Anterior | | Layer # 232 | |
| 10.6a | Dissectible | | Anterior | | Layer # 214 | |

UNIT 11 | *The Respiratory System*

| Atlas Figure | Section | Gender | View/Show Images For | Region/ System/View/ Type/All | Scroll To | Image Title |
|---|---|---|---|---|---|---|
| 11.2 | Dissectible | | Lateral | | Layer # 209 | |
| 11.3 | Atlas | | Region | Head & Neck | | Lateral Wall of Nasal Cavity |
| 11.4 | Atlas | | View | Medial | | Sagittal Section of Head & Neck |
| 11.6a | Atlas | | View | Anterior | | Laryngeal Muscles (Ant) |
| 11.6b | Atlas | | View | Posterior | | Laryngeal Muscles (Post) 2 |
| 11.7 | Atlas | | Region | Head & Neck | | Laryngeal Muscles (Sup) |

*The Respiratory System (continued)*

| Atlas Figure | Section | Gender | View/Show Images For | Region/ System/View/ Type/All | Scroll To | Image Title |
|---|---|---|---|---|---|---|
| 11.8 | Atlas | | System | Respiratory | | Bronchial Tree (Ant) |
| 11.11 | Atlas | | System | Respiratory | | Thoracic Viscera (Ant) |
| 11.12a | 3D | | 3D Lungs | | | Apex of Right Lung |
| 11.12b | 3D | | 3D Lungs | | | Apex of Left Lung |
| 11.12c | Atlas | | View | Medial | | Right Lung (Med) |
| 11.12d | Atlas | | View | Medial | | Left lung (Med) |
| 11.13 | Atlas | | View | Superior | | Mediastinum (Sup) |
| 11.14 | Atlas | | Type | Cadaver Photograph | | Dissection of Left Mediastinum |

*The Digestive System*

| Atlas Figure | Section | Gender | View/Show Images For | Region/ System/View/ Type/All | Scroll To | Image Title |
|---|---|---|---|---|---|---|
| 12.2 | Dissectible | | Anterior | | Layer # 195 | |
| 12.3 | Atlas | | System | Digestive | | Inferior Mesenteric Artery 1 |
| 12.6 | Atlas | | Type | Cadaver Photograph | | Dissection of the Facial Nerve |
| 12.8 | Dissectible | | Anterior | | Layer # 204 | |
| 12.9a | Dissectible | | Anterior | | Layer # 198 | |
| 12.9b | Dissectible | | Anterior | | Layer # 205 | |
| 12.11 | Atlas | | View | Inferior | | Liver (Inf) |
| 12.11b | Atlas | | View | Inferior | | Abdomen at T12 Vertebra (Inf) |
| 12.12a | Atlas | | Region | Abdomen | | Abdominal Viscera |
| 12.13a | Atlas | | Region | Abdomen | | Intestines and Mesentery |

| Atlas Figure | Section | Gender | View/Show Images For | Region/ System/View/ Type/All | Scroll To | Image Title |
|---|---|---|---|---|---|---|
| 13.2 | Dissectible | Male | Anterior | | Layer# 240 | |
| 13.4 | Dissectible | | Anterior | | Layer# 232 | |
| 13.5 | Atlas | | System | Urinary | | Renal Arteries |
| 13.9 | Atlas | Female | View | Medial | | Female Pelvic Organs (Med) 1 |

U N I T  1 4 | *The Reproductive System*

| Atlas Figure | Section | Gender | View/Show Images For | Region/ System/View/ Type/All | Scroll To | Image Title |
|---|---|---|---|---|---|---|
| 14.1a | Atlas | Male | Region | Pelvis & Perineum | | Male Pelvic Organs (Med)1 |
| 14.1b | Atlas | Female | Region | Pelvis & Perineum | | Female Pelvic Organs (Med)1 |
| 14.2 | Atlas | Male | Region | Pelvis & Perineum | | Fascia in Male Pelvis (Med) |
| 14.3a | Atlas | Male | View | Lateral | | Male Pelvic Organs (Lat) |
| 14.6 | Atlas | Male | System | Reproductive | | Blood Supply To Male Pelvis |
| 14.7 | Atlas | Female | System | Reproductive | | Triangles of Perineum (Female) |
| 14.8 | Atlas | Female | Region | Pelvis & Perineum | | Fascia in Female Pelvis (Med) |
| 14.9b | Atlas | Female | View | Medial | | Female Pelvic Organs (Med)2 |
| 14.14 | Dissectible | Female | Lateral | | Layer # 2 | |

## Illustration Credits

The following illustrations are adapted from *Principles of Human Anatomy* 8<sup>th</sup> edition by Gerard Tortora (New York, 1999) John Wiley & Sons ©1999 Biological Sciences Textbooks, Inc.:

1.1, 1.2, 1.3, 1.4, 1.5, 2.1, 2.2, 2.3, 2.4, Table 3.1, 3.2, 3.3, 3.4, 3.6, 3.8, 3.9a-b, 3.10a, 3.11, 3.16, 3.17a-c, 3.18, 3.20, 3.21b, 3.22a, 3.23, 3.24, 3.25, 3.26, 3.27, 3.28, 3.29, 3.30, Table 4.1, 4.4, 4.5, 4.8, 5.1, 5.2, 5.4, 5.5, 5.7, 5.8, 5.11, 5.12, 5.13, 5.16, 5.17, 5.18, 5.19, 6.1, 6.2a-b, 6.3, 6.4, 6.5, 6.7, 6.13, 6.15, 6.17, 6.19, 6.23, 7.1, 7.2, 7.3, 7.4, 8.1, 8.2, 8.3, 8.4, 8.5, 8.8, 8.13, 8.14, 8.16, 8.18, 8.19, 8.21, 8.22, 8.25, 8.26, 9.1b-c, 9.2a-d, 9.3a, 9.3c, 9.4, 10.1, 10.2, 10.3, 10.4, 10.5, 10.6, 11.1, 11.3, 11.4, 11.7, 11.9, 11.10, 11.11, 11.13, 11.14, 11.15, 12.1, 12.5, 12.6, 12.7, 12.8b, 12.9b-c, 12.11, 12.12, 12.13b, 13.1, 13.3, 13.4, 13.5, 13.7, 13.8, 13.9, 14.1, 14.3, 14.4, 14.5, 14.8, 14.9, 14.10, 14.11b, 14.12, 14.13, 14.14a, 15.1, 15.2

The following illustrations are adapted from *Principles of Anatomy and Physiology* 8<sup>th</sup> edition by Gerard Tortora and Sandra Reynolds Grabowski (New York: John Wiley & Sons, 1996) ©1996 Biological Sciences Textbooks, Inc. and Sandra Reynolds Grabowski: 4.2, 9.7, 12.4

3.1a, 3.31 From *Atlas of the Human Skeleton* by Gerard J. Tortora (New York, 1996) John Wiley & Sons.

14.14b From *Introduction to the Human Body*, 4th edition by Gerard J. Tortora ((New York: John Wiley & Sons, 1997) ©1997 Biological Sciences Textbooks, Inc.

## Photo Credits

### Chapter 1:

1.5a-c(1) Stephen A. Kieffer and E. Robert Heitzman, An Atlas of Cross-Sectional Anatomy, Harper & Row, Publishers Inc., New York. 1.5a-c(2) ©Lester Bergman and Associates. 1.5a-c(3) ©Martin Rotker.

### Chapter 2:

2.1a(1) ©Photo Researcher. 2.1a(2) ©Douglas Merrill. 2.1b ©Ed Reschke. 2.1c ©Ed Reschke. 2.1d ©Douglas Merrill. 2.1e ©Biophoto/Photo Researchers. 2.1f ©Biophoto/Photo Researchers. 2.1g ©Biophoto/Photo Researchers. 2.1h ©Andrew J. Kuntzman. 2.1i ©Ed Reschke. 2.1j ©Bruce Iverson. 2.1k ©Lester Bergman and Associates. 2.2a ©R. Kessel/Visuals Unlimited. 2.2b ©Lester Bergman and Associates. 2.2c ©Biophoto/Photo Researchers. 2.2d ©Ed Reschke. 2.2e ©Biophoto/Photo Researchers. 2.2f ©Andrew J. Kuntzman. 2.2g ©Ed Reschke. 2.2h ©Biophoto/Photo Researchers. 2.2i ©Biophoto/Photo Researchers. 2.2j ©Biophoto/Photo Researchers. 2.2k ©Biophoto/Photo Researchers. 2.2l ©Ed Reschke. 2.2m(2) ©Ed Reschke. 2.2m(4) ©Biophoto/Photo Researchers. 2.2m(5) ©Biophoto/Photo Researchers. 2.2m(6) ©Ed Reschke. 2.2m(7) ©Lester Bergman and Associates. 2.2m(8) ©Biophoto/Photo Researchers. 2.2m(9) ©Ed Reschke. 2.3a ©Ed Reschke. 2.3b ©Ed Reschke. 2.3c ©Ed Reschke. 2.4 ©Ed Reschke

### Chapter 3:

3.1a-b, 3.4, 3.5, 3.6, 3.8, 3.9b-c, 3.10b, 3.10c, 3.10d, 3.11, 3.12, 3.15, 3.21a-b, 3.22a-c,3.23a-b, 3.24a-b, 3.25, 3.26a-b, 3.27a-b, 3.28a-b, 3.30a-b, 3.31: Photo by Mark Nielsen.

3.2, 3.3, 3.7, 3.16a-b, 3.17a-c, 3.18a-b ©Mark Nielsen 3.9a, 3.10a, 3.13, 3.19a-b Mark Nielsen

### Chapter 4:

4.1, 4.2, 4.7 Mark Nielsen

4.3, 4.4, 4.5, 4.6, 4.8 © Mark Nielsen

### Chapter 5:

5.4, 5.5, 5.7, 5.8, 5.11, 5.12, 5.13, 5.16,5.17, 5.18, 5.19 © Mark Nielsen

# INDEX

NOTE: A *c* following a page number indicates a cat dissection and an *s* following
a page number indicates a sheep dissection.

243

**245**